CALL YOUR "MUTHA'"

 *Heretical Thought*

HERETICAL THOUGHT

Series editor: Ruth O'Brien,
The Graduate Center, City University of
New York

Assembly
Michael Hardt and Antonio Negri

The Rise of Neoliberal Feminism
Catherine Rottenberg

*Insurgent Universality: An Alternative Legacy of
Modernity*
Massimiliano Tomba

CALL YOUR "MUTHA'"

*A Deliberately Dirty-Minded
Manifesto for the Earth Mother in the
Anthropocene*

JANE CAPUTI

OXFORD
UNIVERSITY PRESS

Oxford University Press is a department of the University of Oxford. It furthers the University's objective of excellence in research, scholarship, and education by publishing worldwide. Oxford is a registered trade mark of Oxford University Press in the UK and certain other countries.

Published in the United States of America by Oxford University Press
198 Madison Avenue, New York, NY 10016, United States of America.

Library of Congress Cataloging-in-Publication Data
Names: Caputi, Jane, author.
Title: Call your 'mutha' : a deliberately dirty-minded manifesto for
the Earth Mother in the Anthropocene / Jane Caputi.
Description: New York, NY : Oxford University Press, [2020] |
Includes bibliographical references and index.
Identifiers: LCCN 2020001343 (print) | LCCN 2020001344 (ebook) |
ISBN 9780190902704 (hardback) | 9780190902711 (paperback) |
ISBN 9780190902735 (epub) | ISBN 9780190902742
Subjects: LCSH: Environmentalism—Social aspects. | Environmentalism—
Philosophy. | Environmental sciences—Language. | Sexism in language. |
Ecofeminism. | Patriarchy. | Nature—Effect of human beings on. |
Geology, Stratigraphic—Anthropocene.
Classification: LCC GE195 .C375 2020 (print) |
LCC GE195 (ebook) | DDC 304.2—dc23
LC record available at https://lccn.loc.gov/2020001343
LC ebook record available at https://lccn.loc.gov/2020001344

To Helene Vann, teacher, friend, and mothering power

CONTENTS

Preface   *ix*

Invocation   1
Introduction: In the Name of the "Mutha'"   3
1.   What's Going On?   7
2.   The Dirty/Earthy Mother   27
    Prelude to a Curse   40
    The Curse   41
3.   The Gods We Worship   42
4.   The Anthropocene Is a Motherfucker   77
5.   Color Mother Nature Gone   110
    Interlude between a Curse and a Swear   142
    The Swear   143
6.   "Feed the Green"   144
7.   "Word Is Born"   181
8.   Call (On) Your "Mutha'"   216
    Convocation   238
    Coda. "Gather and Vote"   239

Poetry Credits   *241*
Acknowledgments   *243*
Notes   *245*
Index   *319*

PREFACE

Mother Earth, a.k.a. Mother Nature, has been called "the oldest religious idea." She has been invoked in ceremonies, discourse, and historical texts from the dawn of recorded language to modern green and/or Indigenous narratives. Today this ancient concept may be more vital than ever, and Jane Caputi's book offers us a sexualized, gendered, and racialized lens to understand what she calls "Mother Nature-Earth." She also shows us how to understand the concept of the Anthropocene, a term not yet twenty years old that encapsulates how humankind has had such an enormous negative impact on the Earth. The Anthropocene manifests in direct contrast to Mother Nature-Earth, is implicitly male instead of female, defined by dominance instead of coexistence, and sustained by destruction instead of reciprocity.

The Anthropocene is not just a geological idea but a theological, philosophical, and political one. It suggests that humankind is now the main element driving the development of the Earth, and it illustrates how the very concept of "progress" has been synonymous with gaining "control" over nature (i.e., raping Mother Nature-Earth) instead of living harmoniously as a *part* of nature, as in the worldview of many Indigenous peoples. When we say something is "down to earth" or "earthy," it is considered to be the opposite of "spiritual," when in fact they should mean the same.

This is not a good thing. Yet not all people are equally responsible for ecocide, the destruction of the ecosystem—an event so far-reaching that there may already be no way back. Indeed, one purpose of the book is to identify the responsible misogynists so that we can understand how the Anthropocene worldview targets the earthy (including particular peoples) as constant victims of violence in this so-called Age of Man (or *Man*, as Caputi pointedly italicizes it).

At the same time, this book expresses devotion to Mother Nature-Earth. Caputi gives us a stunning and graphic array of material exemplifying this devotion, ranging from dreams and visual culture to speculative fiction and poetry. She shows how language, folklore, and popular culture reveal a tradition of resistance. And a large part of that resistance is embodied in what the scholar Geneva Smitherman calls "Black talk," especially in the word "motherfucker" (or "muthafucka").

"Motherfucker" is "The Man's" defining act: The destruction that is the Anthropocene. It is the part in which The Man is the "eater of Others," a role that contains genocide, gynocide, and ecocide. Just as slave masters raped women and claimed ownership of both mother and offspring, the "techno-military-industrial-consumer global complex" rapes Nature-Earth and grabs the profits. "The Anthropocene is, in short," Caputi writes, "*The Age of the Motherfucker.*"

The positive aspect of the book is how Caputi helps us find ways to "un-*Man*" and "Un-*Missy*" by understanding how to discredit certain ruling knowledge systems and credit others. Caputi seeks to practice what Robin Wall Kimmerer explains as "Indigenous ways of knowing"—in the mind, body, emotion, and spirit. This all-inclusive epistemology creates empathy as it highlights intuition and dissolves Euro-Western dualities by emphasizing synchronicities. No more hierarchical differentiation between nature and culture, soil and spirit, dirty and clean, black and white, earth and sky, or female and male; these are all invidious.

What the word "Mutha'" conveys to Caputi is the values of the inexorable autonomy and formidability of Nature-Earth as well as the mutuality, reciprocity, interdependence, and community found in traditions not aligned with *The Man*. Indeed, she quotes the cofounder of Black Lives Matter, who says that climate change is central to the movement because it was caused by a racist system that harmed people of color more than others. Instead of focusing on dualities, this "Mutha'" allows people to "listen in" so they can understand the human capacity for interconnectedness. Mother Nature-Earth is shape-shifting, female, femme, and ambisexual. Another way of putting it: Mother Nature-Earth is energy and matter, essence and change, dark and light, or the solution and the mystery.

The "standard" English language militates against a proper understanding of these concepts, but "spiritual meanings," as Caputi learns from traditional and Indigenous ways of thinking, show how we can return to a more balanced life lived as an integral element of the natural world, instead of something separate from and above it. In this way Mother Nature-Earth is not just a fairy tale, cartoon character, or figure of speech, but the key to saving our planet.

Giving us a way to understand both the destruction and the resurrection of Mother Nature-Earth brings the reader to the edge of a new consciousness with verve and great energy. In this way, Caputi's brilliant analysis pushes us toward heretical thought.

Ruth O'Brien, The Graduate Center,
The City University of New York

INVOCATION

From the Latin "invocāre to call upon . . . to implore; to call by name." (OED)

Call Your "Mutha'". Jane Caputi, Oxford University Press (2020). © Oxford University Press.
DOI: 10.1093/oso/9780190902704.001.0001

When Earth Becomes an "It"

When the people call Earth "Mother,"
they take with love
and with love give back
so that all may live.

When the people call Earth "it,"
they use her
consume her strength.
Then the people die.

Already the sun is hot
out of season.
Our Mother's breast
is going dry.
She is taking all green
into her heart
and will not turn back
until we call her
by her name.

Marilou Awiakta, *Selu: Seeking the Corn-Mother's Wisdom*

INTRODUCTION

In the Name of the "Mutha'"

THIS BOOK TURNS ON THE word *motherfucker*. It is a word that springs from the genius of what Geneva Smitherman calls "Black Talk."[1] and bears a seemingly infinite range of variations— *muthafucka', mothafucka*—and meanings, from the very worst thing you can call someone to the very best and sometimes, even, nothing at all.

The word here is that the *Man* as well as the *Human* in the proposed new geological age—the Anthropocene ("Age of Man" or "Age of Humans")—does not signify everyone of our species. The *Man* behind the Anthropocene is the *motherfucker* in the word's original and worst sense (the most vicious of oppressors and exploiters). At the same time, perennial Mother Nature-Earth epitomizes that word's best sense—the *"Mutha',"* the indomitable, inexhaustible, uncatergorizable, and untranslatable (into official terms of gender and everything else) force-source of continuous birth, growth, death, transformation, and renewal.

The classification of the *Anthropocene* is intended to address the qualitative impact that humans are having on the global eco-system. The present time is characterized by the seemingly boundless reach of technological mastery, but also loss of habitat and early death for many human and other-than-human beings, mass extinction, climate change, environmental migrations as people flee adverse ecological shifts, a new Nuclear Age, pandemics,

Call Your "Mutha'". Jane Caputi, Oxford University Press (2020). © Oxford University Press.
DOI: 10.1093/oso/9780190902704.001.0001

topsoil blowing away forever in the wind, a plastic ocean, wars over oil as well as water, a world without birds and bees, an insect apocalypse, an airpocalypse, and an all-round eco-apocalypse. Despite obvious cause for alarm, Anthropocene enthusiasts and even some critics remain confident that humanity's manifest destiny is to conquer nature, and that these troubling matters can be resolved with the application of a uniquely human ingenuity as well as further penetration into nature's secrets.

There are vital scholarly critiques of the Anthropocene and its arrogance and injustices, particularly its embeddedness in racism, colonialism, and consumer capitalism.[2] There remains a need for recognition of the ways that the Anthropocene is simultaneously the expression of the inherent rapism[3] of a misogynist and heteropatriarchal[4] culture. Too few, even of the era's critics, reckon with the common understanding of ecocide as the *rape* of the Earth. Nor do they face the reality of Mother Nature-Earth and wonder what that original force-source might have to say about all this.

What calls me to this work, then, is twofold. One purpose is to continue the consideration of the related oppressive hatreds and violences, all infused with the fear of (inner and outer) nature,[5] that have led to the Anthropocene. Another is to *invoke* the "Mutha'," to "call on . . . appeal to . . . summon . . . conjure . . . utter (a sacred name)" (*OED*).[6] This conjuration itself participates in the "Mutha's" powers of ending one way and beginning another. Calling the "Mutha'" is to refuse to reproduce the Anthropocene and to create something new.

Accordingly, I rely on sources and directions heretical to the ruling episteme. As I am an American studies scholar, I consult primarily the knowledge brought to bear by cultural theorists, philosophers, scientists, artists, and thinkers from this part of the world, and specifically Indigenous and African diasporic ones. I also attend closely to poetry, song, my own dreams, visual culture, speculative fictions, activist watchwords, and a resistant oral tradition carried by select pockets of folklore, popular culture, and language, including in some words characterized as obscene—notably, *motherfucker.*

That word was born in 19th-century slavery to name the White slave master, according to the African American oral tradition. Here is the *Man* behind the "Age of Man," and *motherfucking* is his defining act. Let's break it down. The word *mother* is manifold, evoking the "source" (*OED*), from the personal to the planetary. The verb *fuck* reveals heteropatriarchy's fusion of sex and violence, holding concomitant definitions of "to engage in sexual intercourse" and to "harm irreparably; finish; victimize."[7] *Motherfucking* is forced sexualized entry into, harm, domination, possession, spirit-breaking, exploitation, extraction, and wasting of another for reasons of power, pleasure, plunder, and profit. *Motherfucking* aims to turn the source into a resource. *Motherfuckers* include rapists, abusers, "pussy" and land grabbers, slave masters, colonizers, conquerors, exploitative governmental leaders, and many of those peopling the highest ranks of militaries and corporations, notably those of fossil fuels and the industrial farming of plant and animal beings. The Anthropocene, in short, is *The Age of the Motherfucker.*

But that is by no means the end of the story. *Motherfucker,* like all living words, moves and evolves. As time passed, the word turned, showing another face and taking on positive meanings. One contributor to the online *Urban Dictionary* defines *motherfucker* as a "formidable and inexorable force."[8] All this can lead to a recognition of Mother Nature-Earth *as* the *"Mutha'"*—the fluid, adamantine, female, femme,[9] non-binary, positive and negative force-source that brings all into being and keeps all going. This "Mutha'" power is strategically *unnamed* and unacknowledged in standard White English. Quiet as *The Man* might try to keep it, the "Mutha'" is the power capable of overwhelming him in the Anthropocene, not the other way around.

As a White, cisgender woman and feminist scholar, I put "Mutha'" in quotation marks to acknowledge my indebtedness to Black Talk and to the speakers, philosophers, artists, theologians, rappers, theorists, linguists, poets, singers, and seers who originally put forth and spin out the word's powers. These include powers of the curse and the swear as well as those of the prayer. These three interrelated powers provide the structure of this book, which begins with *Invocation*, moves into *The Curse*, answers that with *The Swear*, and concludes with *Convocation*.

Raised to be a good Catholic, I was taught to begin an invocation by saying "In the name of the Father"—a piety affirming a world of men over women, up over down, light over dark, thought over feeling, and mind over matter. As befits this work, appearing as it does in a series on Heretical Thought, I deliberately go contrary-wise to pieties both religious and rationalist. I begin and end in the name of the *"Mutha'."*

WHAT'S GOING ON?

THE 2005 *MILLENNIUM ECOSYSTEM ASSESSMENT,* compiled by 1,360 leading scientists from ninety-five countries, reported that anthropogenic activities are damaging the planet at a rapid rate and to the point that the "ability of the planet's ecosystems to sustain future generations can no longer be taken for granted."[1] This and other such reports urge drastic, immediate, and available changes: a major decrease in consumption, including the eating of meat and use of private vehicles; the implementation of sustainable land practices; the local regulation of resources; better education; the use of sensible technologies, including solar and wind power; and the imposition of higher costs on those who exploit ecosystems. Otherwise, it is likely that Earth will no longer so readily be gifting us with purification of air and water, protection from natural disasters, and foods and medicines.

This possibility of eco-apocalypse begs the cosmic question posed by Marvin Gaye and so many others: "What's going on?"[2] A key part of the answer requires bringing two gendered, racialized, and sexualized concepts—*Mother Nature-Earth* and the *Anthropocene*—to the fore.[3]

"Mother Earth" is probably, as religious studies scholar Catharine Roach has determined, "*the* oldest human religious idea";[4] and, I would add, the oldest scientific and philosophical idea. In her *Dictionary of Nature Myths*, Tamra Andrews writes that the earliest human naming of an "Earth God" took the shape of a "mother," who "appeared to embody both the earth and nature itself," the "womb that gives life and the tomb that receives

Call Your "Mutha'". Jane Caputi, Oxford University Press (2020). © Oxford University Press.
DOI: 10.1093/oso/9780190902704.001.0001

the dead . . . a single animating power . . . the process of cosmic renewal."[5] Even today, Mother Nature-Earth's divine character sometimes surfaces in the mainstream, though it is often met with antagonism and rivalry from patriarchal religions. Pope Francis, remarkably, quotes Francis of Assisi referring to "our Sister, Mother Earth," chastises those who believe themselves "entitled to plunder her at will," and urges ecological sanity.[6] But still, the Earth is not acknowledged as divine, as the creator. And in October of 2015, the Church clarified its position on cremation in a statement confirming that the death of the body must be understood as a preliminary stage before ultimate reunification with the male father God in heaven and not "the moment of fusion with Mother Nature . . . or as a stage in the cycle of regeneration."[7]

A European worldview based in love and respect for traditional Mother Earth prevailed to some degree prior to the 16th century, historian Carolyn Merchant writes, and constrained some ecologically damaging activities, notably mining, which was understood as the "rape or commercial exploitation of the earth."[8] During the ensuing European scientific revolution, however, this understanding was systematically undermined by the conjoined efforts of patriarchal theology, philosophy, and science, turning self-ruling Mother Earth into a feminized, passive environment and then a subordinated, violable, rapeable, and exploitable resource. Nowadays, Mother Nature-Earth's name continues to be heard in the Eurocentric world, but usually only in ways that make her an unreal and folkloric figure from a defunct, if colorful, oral tradition. A range of caricatures stamp her as the unreasonable bitch, the bringer of bad weather, an annoying nag, a ridiculous frump, or a dumped "whore," on the one hand, or an all-loving, giving, forgiving, and "happy slave," on the other.[9] As such a stock figure, she is used to sell just about anything— margarine, laundry products, pesticides, sugary drinks, and gas guzzlers.[10]

Mother Nature-Earth remains a significant force, though, in many spiritual, scientific, and political traditions outside the

Eurocentric. The Indigenous-led International Movement for Mother Earth Rights defines Mother Earth as an "indivisible, living community of interrelated and interdependent beings with a common destiny . . . the source of life, nourishment and learning." The movement points to the harms of "the capitalist system," including "all forms of depredation, exploitation, abuse and contamination [that] have caused great destruction, degradation and disruption of Mother Earth, putting life as we know it today at risk through phenomena such as climate change."[11] This being of Mother Nature-Earth is real, calling all to a concomitant realization of individual and group potency, growth, and transformative action.

The second concept on the table is the *Anthropocene*. First used in the early 2000s, the term is derived from the Greek *anthropos* and means the "Age of Man" or "Age of Humans." Science writer Elizabeth Kolbert describes the Anthropocene as a "new name for a new geologic epoch—one defined by our own massive impact on the planet.[12] As of this writing, the designation has not yet been officially accepted by all the international geological associations, but seems to be moving in that direction.

While many of those who use the term express the gravest of concerns over human planetary impacts, other Anthropocene enthusiasts predict instead the coming glory of "a new Earth sculpted by humans,"[13] one fulfilling a long-standing goal for the species to become "the masters of planet Earth."[14] They celebrate the current moment as an unprecedented time when "human activity" has become a "globally potent biogeophysical force,"[15] and *Homo sapiens* are poised to "remake nature according to their wishes,"[16] even to become a "God species."[17] Some openly wonder if "humans" now are on the verge of "overwhelming the great forces of Nature."[18] As this sort of discourse suggests, the Anthropocene is nominally a geological idea but also a political (based in power relations), philosophical (concerning the nature of existence), and theological (who gets worshipped) one.

These interlocking meanings of the Anthropocene, a.k.a. "Age of Humans," cohere in the notion of the "human" it enthrones. Decolonial theorist Sylvia Wynter gets to the heart of the matter, observing that the word *human* doesn't now and hasn't ever really signified everyone. She uses the term *Man* to signify a ruling "ethnoclass," typically Western, White, male and male-identified, masculinist, straight, professional, affluent, conventionally abled, and socially powerful. *Man*, Wynter perceives, claims the mantle of civilization (rationality, culture, knowledge, and progress) and "overrepresents itself as if it were the human itself."[19] The Black English expression "The Man" operates along similar lines, as it designates a "policeman" or "any white authority figure."[20] This is not a biological destiny. Attaining entry into the club of *Man* is not guaranteed simply by sex or skin tone. The membership excludes some potential members because of their difference from the oppressive norm and/or their refusal to uphold it. That *Man* club also welcomes and relies upon certain complicit others, including well-placed White women whom we might think of as *Missy*.[21]

In *Man's* political-philosophical-theological system, the guiding precept is human exceptionalism, the notion that humans are a superior species, ordained to conquer "nature." This founds *speciesism*, defined by philosopher Lisa Kemmerer as "institutional power and authority . . . used to support and perpetuate the oppression of nonhuman individuals,"[22] while also legitimating the oppression of some humans, who are, as feminist sociologist Maria Mies puts it, "'defined into nature.'"[23] Ecowomanist thinker Shamara Shantu Riley explains: "The global environmental crisis is related to the sociopolitical systems of fear and hatred of all that is natural, nonwhite, and female that has pervaded dominant Western thought for centuries."[24] The list of those whom *Man* excludes from the human is long and includes not only women (which always includes trans women), lesbians, gay men, trans men; bisexual, intersex,[25] and non-binary people; ethnically/racially othered people; disabled people; Indigenous

2SQ ("Two Spirit and queer")[26] people; and all others categorized, variously, as inherently freakish, marginal, monstrous, savage, contaminating, dependent, and unassimilable.

The notion of an "Age of Humans" suggests, moreover, that is just "human nature" to dominate. It also, just as deceptively, identifies the entire species as the responsible party. Writer Toni Cade Bambara calls out the actual culprits, asking: "Can the planet be rescued from the psychopaths?" Bambara is not using that loaded word lightly. She means those people who by virtue of race, sex, and class privilege name their own power-mad, egocentric, and greedy selves as, not only the norm, but also the smartest, the very best civilization has to offer, the ones rightly at the center and on top. At the same time, such "psychopaths" grant themselves the right "to define, deform and dominate"[27] everyone else.

As the humans most culpable for ecocide (devastation of an ecosystem beyond the possibility of recovery) profit from it, at least in the short run, the ones least responsible for it suffer the loss of homes and livelihoods; toxic workplaces; illness and premature death;[28] anxiety, depression and despair; increased risk of sexual violence; induced sterility; poisoned breast milk; imprisonment; wars over resources; and "land trauma"—"embodied feelings of breeched consent over lands and bodies."[29] The common critical Anthropocene assumption that "humans" might cause the death of the planet, theorist Elizabeth Povinelli aptly points out, works to cover up the reality that *some* humans are orchestrating the "letting die or killing of specific human populations."[30]

Non-human beings are suffering grievously as well. Anthropocene activity and effects are decimating wild animal beings, like birds.[31] Meanwhile, human demand has made domestic chicken beings the most common birds on the planet, and as a species they are even said to "define the Anthropocene."[32] Billions of genetically modified chickens are bred industrially and treated in torturous ways, while the humans who labor to process them into products are poorly paid and trapped in dangerous jobs.[33] Chickens' bones have now accumulated to such an extent in the fossil record that some geologists say that they—along with radioactive waste and plastic trash fused with rocks—provide geophysical evidence for the Anthropocene.[34]

Some analysts find the term *Anthropocene* both obscuring and arrogant. Humanities scholar Jeremy Davies uses it in his *Birth of the Anthropocene* but indicates that his interest is only in the stratigraphic meaning of the era and not the self-serving narrative about nature now being "completely subordinated to human agency."[35] The editors of the insightful anthology *Arts of Living on a Damaged Planet* also use the word *Anthropocene*, but reject those "worst uses of the term, from green capitalism to technopositivist hubris."[36] These self-serving uses of the term *Anthropocene*, though, are my focus precisely because they most reveal the conjoined political, philosophical, and theological meanings of the era, as well as its foundations in rapist and racist violence.

MAN OVERWHELMS NATURE

A TV commercial for the Apple iPhone,[37] from 2019, shows a series of images of "nature"—typically rendered in a voyeuristic manner as spectacular, charismatic, and outside the human. We see eagles, glaciers, volcanoes, Venus flytraps, elephants, lightning, avalanches, and so on. There is never a human in sight. The images of spectacular nature proceed in greatly accelerating flashes accompanied by piano music that starts off gently, turns dirge-like, and ultimately shifts to the exuberant, blaring guitars of hard (White and masculine-identified) rock. A title appears with the admonition "Don't mess with Mother," followed by a spitting lizard. The soundtrack concludes with a triumphant male vocalization and a final ejaculation from the guitars. The final and telling title card reads, "Earth shot by iPhone." Susan Sontag's perception that taking a photograph can be a means to "violate," to turn others "into objects that can be symbolically possessed"[38] is apt here. The iPhone, despite the supposed warning, is actually promising its wielders increased power to "mess with Mother."

Chemist and Nobel laureate Paul Crutzen, credited with first bringing attention to the Anthropocene, coauthored an essay with chemist Will Steffen and historian John McNeill titled "The Anthropocene: Are Humans Now Overwhelming the Great Forces of Nature?" The authors speak of the era as "an unprecedented time," when "human activities have become so pervasive and profound that they rival the great forces of Nature and are pushing the Earth into planetary *terra incognito* . . . less biologically diverse, less forested, much warmer, and probably wetter and stormier."[39] The alarm raised is sound and loud, but the words used—*human, rival, overwhelm, nature*—need some scrutiny, as they carry the same troubling conflation of admonition, admiration, and arrogance, as well as a trinity of problematic political, philosophical, and theological assumptions.

The word *human*, again, masks the political reality that those who are responsible for and benefit from the Anthropocene are not all humans, but specifically the ones standing atop an intellectual-techno-military-industrial-consumer global complex. The concept of *rivalry* is inseparable from the emotion of *envy*, "the feeling of mortification and ill-will occasioned by the contemplation of superior advantages possessed by another" (*OED*). Sigmund Freud famously opined that women were victims of "penis envy,"[40] which psychoanalyst Karen Horney countered with an alternative theory of men's envy of the womb's procreative and creative powers.[41] *Man's* womb envy manifests macrocosmically in his assault against the primal womb-tomb who is Nature-Earth. An exemplary ad for the Chevrolet Captiva SUV makes this plain. It shows the vehicle (itself a symbolic womb for a fantasy fetus-driver to now steer and control) in a showroom, digitally mastered to make it seem large enough to tower over a minimized Arctic terrain. Bear, wolf, pine tree, even snow-peaked mountain beings are all teeny-tiny next to the colossal Captiva. The copy promises that the vehicle will alleviate mortification: "Don't let nature make you feel insignificant."[42]

The second verb at issue is *overwhelm*, which means to "overturn, overthrow, upset; to turn upside down . . . to bring to ruin, to destroy utterly" (*OED*). *Overwhelm*, not surprisingly, carries profound resonances of sexual violence and racial violence. Ecofeminist and ecowomanist thinkers and activists have for decades shown how men's gender-based sexual violence against women (including cis and trans women) is linked inextricably with violence against Nature-Earth.[43] *Overwhelm* itself can be synonymous with rape. Social psychologist Diana E. H. Russell, in *The Politics of Rape*, speaks of men who use their social power to "overwhelm women."[44]

Scholar Christina Sharpe, in her potent and meditative *In the Wake: On Blackness and Being*, asserts that to understand Black people living "in the wake" of slavery is "to recognize the ways that we are constituted through and by continued vulnerability to overwhelming force though not *only* known to ourselves and to each other *by* that force."[45] The word *overwhelm* thus reveals some of the ways that this sexualized, gendered, and racialized domination resides at the very foundation of the Anthropocene, along with the assumption that those being overwhelmed are inherently unable to push back.

The object of Anthropocene *Man's* rivalry is *nature*, a term also offering fruitful opportunities for insight. A mothering power is implicit in the word *nature*, which comes from a Latin root meaning "birth . . . genitals, creative power governing the world" (*OED*). That etymology reflects an understanding of nature as a process of ongoing birth or renewal. Public intellectual and activist Winona LaDuke (Ojibwe)* refers to *"mino bimaatisiiwin"*— a word in the "Anishinabé (Chippewa)" language, alternately translated as "the good life" or "continuous rebirth."[46] A similar

---

* I give the tribal affiliation the first time I cite an Indigenous person.

understanding of life was recognized in older European concepts of *natura naturans*—nature naturing, or actively making life—opposed by a competing notion of *natura naturata*, the manifest "creation" made by an external male God.[47]

In a speech at a college graduation in 1962, Rachel Carson—the environmentalist, biologist, and author of the classic *Silent Spring*[48]—succinctly defined *nature* as "the part of the world that man did not make."[49] Carson is adamant, though, that humans are "part of nature" and refuse to acknowledge that at their peril: "Man has long talked somewhat arrogantly about the conquest of nature; now he has the power to achieve his boast. It is our misfortune—it may well be our final tragedy—that this power has not been tempered with wisdom, but has been marked by irresponsibility; that there is all too little awareness that man is *part* of nature, and that the price of conquest may well be the destruction of man himself."[50] The mass-murderous and simultaneously suicidal Anthropocenic assault on the planet is reflected in what is nowadays a common crime in the United States: an enraged individual, most often a Euro-American man, commits mass murder with a high-powered rifle and either kills himself or is killed by the police at the scene.[51] Such killers more often than not have a history of expressing misogyny and abusing their wives and other members of their families. They also often espouse racist animosities, including ones specifically directed against immigrants and refugees in a time of increasing migration due to ecological breakdowns. Journalist Naomi Klein perceives such mass shootings as ideologically linked to a "white power eco-fascism" that is emerging from the "wealthiest strata of society" who have contributed most to climate change, but are determined to avoid their "responsibilities."[52] Allusions to a masterful White-identified masculine potency—a hero or god "messing with," probing, peering at, struggling against, and ultimately overwhelming a dark, chthonic Earth-Mother—appear regularly in Western scientific, philosophical, theological, and popular traditions.[53] These climax in much Anthropocene discourse with triumphant *Man* taking on the mantle, not only of that supposed hero, but also of *God*. Science writer Mark Lynas boastfully titles his tome on the Anthropocene *The God Species: Saving the Planet in the Age of Humans*.[54] Does it even need saying that the "God" providing the divine role model here is *Man's* version of divinity, made in his own egocentric image—the White, purely male, heavenly patriarch, supposedly perfect, controlled and controlling, immortal, omniscient, omnipotent, and limitless?

Lesbian-feminist theologian Mary Daly unforgettably argues that patriarchal men have made this "God" in their own image to justify male supremacy. In *Beyond God the Father* (1973), she writes that the figure of "God in 'his' heaven" as a "father ruling 'his' people," makes it seem that it is "in the 'nature' of things and according to divine plan and the order of the universe that society be male-dominated."[55] *Man* plays this *God*, doing his damnedest to lord it over everyone else, including elements and plant and animal beings, while also trying to commit deicide against his *rival*—Mother Nature-Earth.

The notion of *heaven* as the home of the divine has further troubling implications. Poet and essayist Linda Hogan (Chickasaw) perceives that "the Western belief that God lives apart from earth is one that has taken us toward collective destruction. It is a belief narrow enough to forget the value of matter, the very thing that soul inhabits. It has created a people who neglect to care for the land for the future generations."[56] This desperate devaluation of matter denies, among other things, the utter dependence of all on the favor of Mother Nature-Earth and the obligation of all to return the favor.

THE GIFTS OF THE EARTH

A troubling series of ads for Tampax tampons turned on the theme "outsmart Mother Nature."[57] In these, Mother Nature, played by a middle-aged Euro-American woman in a dowdy green suit, attempts to deliver a red-wrapped gift package (menstruation) to a young woman. Various gals are shown to quite smartly spurn and even obliterate the gift. One of the print ads depicts a hapless Mother Nature as a teeny-tiny unhappy figure standing next to a gigantic plastic tampon applicator. The copy commands: "Cut Mother Nature down to size." This script turns on the overt belittlement of the original divinity, Mother Nature, implicitly directing worship to the outsized artifact. This is a prescription for disaster.

The colors in play—mostly plant green with a dash of periodic red—bespeak the gift of life, with an ongoing cycle that necessarily encompasses birth, growth, decline, death, and renewal.[58] Plant biologist Robin Wall Kimmerer (Potawatomi) speaks of the ways that human beings are "showered every day with the gifts" of Mother Nature-Earth. This generosity obliges both recognition and reciprocity: "I've been told that my Potawatomi ancestors taught that the job of a human person is to learn, 'What can I give in return for the gifts of the Earth?' "[59] The supreme folly of these Tampax ads is an arrogant refusal of the gift, obviating, too, the requirement to return the favor. This directive should occasion a state of red alert. If Mother Nature actually were to go away, taking her gifts with her, it would be the end of life as we know it.

The belittlement of Mother Nature-Earth by *Man* is a given. Still, the mothering power remains large, serious, and reverenced elsewhere, including overtly or implicitly in social justice movements. At the beginning of the 20th century, feminist and anarchist firebrands Emma Goldman and Max Baginski named their publication *Mother Earth*, hailing Mother Earth as a beacon able to "attract and appeal to all those who oppose encroachment on public and individual life." Goldman and Baginski advocate working for "a new dawn for a humanity free from the dread of want, the dread of starvation in the face of mountains of riches."[60] Mother Nature-Earth also is invoked in a spiritual-political-scientific-legal context in the body of Indigenous knowledge that Greg Cajete (Kha'p'oo Owinge, Santa Clara Pueblo) calls "Native science."[61] Legal scholar Leroy Little Bear (Blackfoot) explains in his preface to that work: "The land is a very important referent in the Native American mind. Events, patterns, cycles, and happenings occur at certain places . . . All of this happens on the Earth . . . so sacred that it is referred to as 'Mother,' the source of life."[62]

Scholars holding to Eurocentric modes might nod politely upon hearing this, but their creed demands its dismissal as, variously, quaint, irrational, or simplistic. Feminists among them might deem it *essentialist*, something falsely represented as immutable and/or universal. This happens, for example, when *Man* claims himself to be "the human" or when "the mother" is put forth as purely and only female, or solely identified as a birthing functionality.

Political theorist Rachel Cochrane reports on Latin American feminists who reject "essentializing" definitions of *Pachamama* that reduce her to gendered caregiver and reproducing uterus. Instead, they name *Pachamama* as a "'whole that goes beyond visible nature.'"[63] In other words, *Pachamama* is matter and energy, the "deity associated with the fertility of the crops," identified "with the fields themselves" but equally "with the power that makes them fruitful."[64] This holistic understanding knows *Pachamama* as both creative force and the manifest world.

These Indigenous conceptions, though, are difficult to convey in the spiritually deficient and gender-polarized language that is standard English. Scholar Alex Wilson (Opaskwayak Cree Nation) elicits another way of understanding: "We call the moon grandmother and the earth mother in English, but in Cree this is not the case. What is important is the relational aspect acknowledging some kind of kinship. In Cree, the land (*aski*) is not gendered."[65] Decolonial theorist Linda Tuhiwai Smith (Ngati Awa and Ngati Porou) speaks of an Indigenous understanding of an "earth parent,"[66] difficult to grasp in an English language mired in false oppositions. She emphasizes that this "earth parent" is not some "mystical, misty-eyed discourse." Indigenous peoples" had to know to survive. We had to work out ways of knowing, we had to predict, to learn and reflect, we had to preserve and protect, we had to defend and attack, we had to be mobile, we had to have social systems which enabled us to do these things. We still have to do these things."[67] Survival on their own land, though, is now impossible for many Indigenous peoples because of settler colonialism, industrial ravaging, and climate change. The *New York Times* reports, for example, on the Uro-Murato people in what is now Bolivia, who have been made homeless as climate change coupled with capitalist-industrial water diversion have dried up and killed Lake Poopó, upon whom the Uro-Murato depended for identity, environs, and livelihood. One fisherman, Adrián Quispe, laments: "The lake was our mother and our father . . . Without this lake, where do we go?"[68]

The Movement for Mother Earth Rights makes it clear that along with ending direct ecocidal destructions and the massive consumption practiced by the affluent, there needs to be a renewed understanding of the reality, intelligence, and complexity of Mother Nature-Earth. This might occasion not only a practical, but also the emotional response suggested by the quintessential green slogan "Love your Mother Earth"[69]—a true love—heartfelt, reciprocal, nonabusive, and unconditional.

NO MORE *MAN*-TALK

Lavinia White, a Haida elder, makes this demand: "*Lets quit talking like men* . . . that's man's talk . . . power . . . They've trashed the lands, they've trashed the oceans, they've trashed the rivers, and now the air . . . it must come to an end . . . if we don't put a stop to it . . . Mother Nature will."[70]

All too often, whether a commentator is celebrating or excoriating the Anthropocene, the frame remains mostly *Man*-talk—Eurocentric theory and related values, often associated with towering intellectual father figures. However valuable

some of this is, cognitive justice[71] demands recognition that
Western knowledge and heritage are not superior, must not be
centered, and, moreover, that these foundations have provided
the political, religious, and philosophical justifications and
economic framework for the Anthropocene. What is needed
now are ideas and actions that have the capacity to un-*Man*,
which requires seeking leadership and guidance elsewhere.
This includes intellectual mother figures and/or thinker-
activists representing what Toni Morrison calls "discredited"
knowledge systems and epistemologies. Morrison describes
these knowledges as discredited because the people who generate them have been and are still being dismissed, eliminated,
marginalized, and ripped off by those enforcing their own
position at the center.[72]

Cognitive justice also demands giving credit where credit is
due. Sociologist Zoe Todd (Métis/otipemisiw) points to an enthu-
siastic embrace by some "Euro-Western" academics of ideas core
to Indigenous ontologies, including the sentience and agency of
Earth and what is now called the "more-than-human world." But,
Todd says, this trend continues the ravages of settler colonialism
because these ideas are recognized only when "espoused by Euro-
Western intellectuals." Said intellectuals do not acknowledge the
Indigenous thinkers, past and present, who have generated the
philosophies and put them into practice in art, law, ethics, and
politics.[73]

Part of my task here is to try to un-*Man* and un-*Missy* myself by,
for example, acknowledging and resisting privilege, taking responsibility, and trying to educate myself and others. This also means
crediting those thinkers and creators who have been unjustly discredited. Accordingly, I do a lot of quoting of those whose knowledge flows into the synthesis that is this book. These include funk
theorists,[74] Black feminists and womanists, Indigenous scientists
and philosophers, *panochonas*,[75] self-named witches,[76] spiritual
activists,[77] poets, artists, Afrofuturists, eco-ability[78] theorists,
revolutionary mothers,[79] Lesbian feminists, LGBTQ thinkers,

ecofeminists, ecowomanists, ecosexuals, and environmental-justice knowledge holders and practitioners.

Such perceiver-teachers draw upon an epistemology very different from the normative positivism,[80] which excludes any way of knowing other than rationality. They suffuse their work with elements all at once personal, cultural, spiritual, artistic, and scholarly. Kimmerer explains that she practices "indigenous ways of knowing . . . [in which] a thing cannot be understood until it is known by all four aspects of our being: mind, body, emotion, and spirit."[81] This includes cultivating empathy, honoring relationship, attending to intuition and meaningful synchronicities,[82] listening to ancestors, learning from other-than-human beings (if they are willing to teach you), hearing the call of land, and heeding and sharing what Emily Brontë calls "queer" dreams.[83] Such distinctive dreams provide, Cajete explains, "natural means for accessing knowledge and establishing relationship to the world."[84] I attend to these alternative ways of knowing, including consulting some of my own dreams, in order to surface an otherwise unavailable awareness about what's going on. The ensuing answer (however partial) is not something I am making up but instead am hearing from others, at times by putting my ear to the ground—and this is not a metaphor.

Language, poet Simon J. Ortiz (Acoma Pueblo) instructs, actually does issue from the ground: "The nature of language . . . has to do with an intuitive and vital connection that human beings have with the natural earth process or dynamic . . . Indigenous people speak language that directly comes from the natural forces all around us . . . Simply put, human language comes from Mother Earth."[85] Irish poet Pádraig Ó Tuama hails the import to him of "the language that came from the earth of this Ireland."[86]

Though *Man*-talk makes himself the source of the word, as well as the only one who speaks, this isn't so. Humans, a very young species, learn how to participate in that process of making life, as well as to think, dream, and speak from other-than-human beings; we are part of the active, intelligent, dreaming, evolving, and speaking community with a shared destiny that *is* Mother Nature-Earth.

THE COMMUNAL AND THE COMMONPLACE

The historian and cultural theorist Vine Deloria, Jr. (Standing Rock Sioux) describes the "communal" character of Indigenous knowledge systems, ones that carry the "distilled experiences of the community" over time.[87] These knowledge traditions come from "a real apprehension of and appreciation for the sacredness of land [which] . . . for traditional peoples, includes the other forms of life that share places with us."[88] These living bodies of Indigenous knowledge, contra colonialist stereotypes, are neither uniform nor static. They emerge from the enormous range of cultures, places, and languages, growing and changing in the context of evolving social, material, and spiritual conditions—including the dire ones of the Anthropocene.

Writer and folklorist Zora Neale Hurston theorizes communal knowledge and alternative epistemologies in African American traditions and arrives at conclusions similar to Deloria's. Hurston speaks of the Black oral tradition, or "folklore," as the "the boiled-down juice of human living," the practical, spiritual, and philosophical knowledge people make of "the natural laws" they encounter in everyday living on and with the land. "The group mind," Hurston continues, "uses up a great part of its life span trying to ask infinity some questions about what is going on around its doorstep." Every generation or so, she continues, someone with "extra-keen perception" (and, no doubt, someone who has asked politely) suddenly grasps something of the "obvious about us." Revealed is a new "law" of the "commonplace" that helps the group "bend" and work with what they find around their own "doorstep," not only to survive, but also to be inventive and avoid boredom.[89] All benefit when infinity offers up at least a few partial answers to the whirling questions regarding the nature of existence, which are then infused into that distilled "juice" of the oral tradition.

Throughout this book, I consult multiple past and contemporary "extra-keen" perceivers, all dwelling on the doorstep of the ultimate commonplace, Earth. All have called on infinity and gotten something of an answer to help understand what's now going on. The hope is to be able to better bend—and bend with—the hurricane-force winds of change now happening.

"SPIRITUAL MEANINGS"

The quest to find out what's going on in the Anthropocene must necessarily recognize the spiritual-energetic meanings of events. An unsigned essay on the White Earth Nation (Chippewa) website points out that Euro-American historians often proceeded as if history did not even happen here until Europeans "visited." Chippewa history, though, records the nation's "full, rich culture"—accessed through written records and archeology but also the oral tradition: "In Chippewa societies, assigned story tellers . . . tell how the earth came to its present form and how people share the earth with all living things. They believe that knowing the spiritual meaning of events is more important than knowing exactly when things happened."[90] *Spiritual meanings* recognize that matter and energy (spirit) are inextricable and that humans are not the only significant dwellers on the land. Greg Cajete explains that in Indigenous knowledge systems "everything is considered to be 'alive' or animate and imbued with 'spirit' or energy . . . connected in dynamic, interactive, and mutually reciprocal relationships."[91]

Historian and theorist Jack Forbes (Powhatan-Renápe, Delaware-Lenápe) writes that though Indigenous systems are diverse, they share commonalities. These include recognizing the "creative process of the universe as a form of thought or mental process," as "plural," and as "seldom pictured as human."[92] He further highlights that Indigenous people "seem to see all of life, which surrounds us, as intelligent, inventive, changing, learning, teaching, evolving, acting, praying, feeling and responding."[93] These understandings qualitatively challenge those that reproduce the Anthropocene.

Journalist Andrew Revkin is typical of the era's enthusiasts in denying consciousness to those who are not human: "Two billion years ago, cyanobacteria oxygenated the atmosphere and powerfully disrupted life on Earth . . . But they didn't know it. We're the first species that's become a planet-scale influence and is aware of that reality. That's what distinguishes us."[94] Perhaps, though, it is Revkin who is unaware, dangerously so because this dismissal of the awareness of others works everywhere to justify their exploitation.

American studies scholar John Mohawk (Seneca) considers the consciousness of other-than-human beings in an ecological context. He explains that the consciousness of a plant is "manifested in its compounds and in its shape at that moment."[95] When human beings, he continues, are able to "acquire" the consciousness of non-human beings, "it becomes extremely difficult to rationalize pollution." Mohawk stresses the need for a return to humanity's "spirit to spirit"[96] relationships with the plants, animals, fibers, waters, and soils upon which our existence depends. He sums up the effects of the standard mode of treating these beings as mere "commodities": "If there's an essence to the world's great problem, that's it."[97]

Jack Forbes stresses one more commonality among Indigenous systems that is particularly pertinent to understanding what is going on in the Anthropocene. This is realizing that "the visible universe" is "not a passive, acted upon, place where only 'immutable laws' of science operate. Instead, its essence seems to be an ability to modify itself in response to new situations."[98] *Man's* Anthropocene story tells how "humanity" has overtaken a mostly silent, passive, speechless, and mindless Earth. It climaxes with "humanity" assuming its rightful place as the "dominant agent of global change."[99] IBM assured the world in an extensive ad campaign that it was making the (supposedly not so bright) Earth into a "smarter planet."[100] All this continues the Western heteropatriarchal tradition of identifying *the human* with itself and as superior to, outside, and over and against nature. It also extends the conceit that said humans are the only ones telling and starring in the drama of the changing Earth. These beliefs are sorely ignorant. The stories that are able to reveal meanings, including spiritual ones, about what's going on in the Anthropocene emerge from the land and speak to the agency of land; they are heard, understood, interpreted, and told by humans, but are not theirs alone.

Speakers of standard (White) English, including me, might be hard pressed to comprehend *spirit* and *spiritual* in this energetic, connective, and earthy context. We have been trained to think of *spiritual* as the antithesis of *earthy* and *matter*. Consulting Webster's unabridged dictionary,[101] one finds that *spiritual* means "incorporeal," and is the very antonym of *earthy*. Correspondingly, *earthy* means "not predominantly spiritual, ideal or ethereal." The related word *earthbound* is defined as wholly "lacking in spiritual quality." In rebuttal, I use *earthbound* to refer to all terrestrial beings energetically and materially bonded in that shared destiny that *is* Mother Nature-Earth.

French sociologist Bruno Latour also uses the word *earthbound* ecologically;[102] but maybe the first to do so was Black feminist scholar bell hooks, in her 2002 essay "Earthbound: On Solid Ground." The standard split between matter and spirit, hooks notes, is a falsity that serves White supremacy and "corrupt capitalism and hedonistic consumerism." She upholds the knowledge of her "sharecropping granddaddy" and other "'backwoods' folks" who taught her that "ultimately, nature rules." There is a knowledge passed down through generations, hooks continues, an awareness that there is "spirit" in nature that must be trusted, that people should "tend the earth," live in "harmony and union with nature," and embrace "the organic rights of the earth."[103]

Monica Sjöö and Barbara Mor, in their classic *The Great Cosmic Mother* (written primarily by Mor while she was a single mother of two on welfare), diagnosed the crisis that is the Anthropocene before the term was around: "By abstracting 'spirit' from 'flesh,' by utterly demonizing the physical world and woman as its source, patriarchy has almost destroyed the home of spirit, which is sacred earth and all flesh."[104] It is imperative to flip *The Man's* script[105]—to acknowledge the Earth as quintessentially spiritual and to attend to the spiritual-energetic relationship of the earthbound. This awareness has since time immemorial been carried by the oral tradition. This is the necessarily dirty-minded story of Mother Nature-Earth.

THE DIRTY/EARTHY MOTHER

POET, ESSAYIST, AND NOVELIST ALICE Walker writes that "by the age of eleven," she knew herself to be a "worshipper of nature."[1] In an essay explaining her turn away from the White Christian God of her childhood, Walker warns: "It is fatal to love a God who does not love you. . . . All people deserve to worship a God who also worships them. A God that made them, and likes them. That's why Nature, Mother Earth, is such a good choice."[2]

Along with Walker and many others, I offer reverence to the Earth Mother, the self-fecundating, endogenous, light-bearing, and intelligent womb-tomb. As further apostasy, I venture beyond the euphemistic "womb" to name the terrestrial power in downright earthy, even "dirty," terms. I invoke *cunt*—signifying the whole bloody bundle.

Cunt can be hurtful to hear. Readers may balk and ask why my language needs to be so downright *dirty*. My intention is not to shock or pain, but to reveal and work with the significant ecological implications of *cunt* and *dirt* itself. Eric Partridge calls *cunt* "the most notable of all vulgarisms," but tells us that it was not always an obscenity. In *A Dictionary of Slang and Unconventional English*, Partridge notes that *cunt* was once "a true language word" (not slang), but "owing to its powerful sexuality the term has since the C15th been avoided in written and polite English . . . and been held to be obscene, i.e. a legal offence to print it in full."[3]

Cunt, for many, rivals *motherfucker* as the most obscene word in the English language, and polite society would silence each of them.[4] Both words, though, have much to say about *Man's* project

Call Your "Mutha'". Jane Caputi, Oxford University Press (2020). © Oxford University Press.
DOI: 10.1093/oso/9780190902704.001.0001

of what Tracy Chapman calls out as the "rape of the world"[5]—subordinating, silencing, and violating women and all whom *Man* defines into nature. *Cunt* became taboo around the same time that European men were attacking both literal and associative cunts. In the context of branding women as witches, the Church sought to enforce the most profound misogyny, enslaving women to procreation by veritably demonizing women for practicing birth control and abortion.[6] Church and state colluded in men's torturing, raping, and executing predominantly women for the heresy of witchcraft. The most influential witch-hunting and torture manual, written by two priests, made it clear that the sin/crime of being a witch was based in "carnal lust, which is in women insatiable."[7] This systematic *gynocide* (misogynist mass murder of women by men) lasted for three centuries.[8] Meanwhile, the Catholic Church officially sanctioned the destruction of sacred groves, the arboreal abode of purportedly obscene nature spirits and places where naked witches met at night to conduct "filthy" rites.[9]

Coincident with *cunt* becoming obscene was the transition from a feudal to a capitalist system, with commons enclosed and peasants forcibly dispossessed, displaced, and converted into cheap wage workers as their rebellions were put down.[10] The economic transformation continued with imperial European landgrabbing, rapist and slave-trafficking invasions, and occupations and the settling of Africa and the Americas. These were justified by an ideology of a male and racial supremacy,[11] and these characterizations were often based on what would benefit the ruling class (the British, for example, would also deem the Irish "savage," suitable to be starved out, their land and goods seized).[12]

These depredations were fueled by what Daly calls "phallic lust . . . a fusion of obsession and aggression [that] rapes, dismembers, and kills women and all living things within its reach."[13] Such lust was epitomized by the gold-obsessed, rape-approving, slave-taking, and mass-murdering Christopher Columbus.[14] Phallic lust shaped what scholar Anne McClintock calls a colonialist imagination of a "porno-tropics," in which manly European conquerors

invade and defeat a people and land cast as negatively dark, matriarchal, sexually excessive, and freakish and defines dark women
as "the epitome of sexual aberration and excess."[15] Phallic lust
also shaped what Deborah Miranda calls "gendercide"[16]—the systemic elimination of feminine, femme, and/or non-binary, queer,
and two spirit people. Meanwhile, the self-appointed civilizers
enslaved, raped, mutilated, and murdered humans and separated
children from their parents, all while systematically deforesting
and wiping out other species, such as the buffalo, to starve out
people, break their spirits, and further effect the conquest.[17]

Finally, also occurring during this period was the scientific
revolution, which had its own misogynist imagination. Merchant
carefully demonstrates how Mother Nature-Earth was being set
up, most vividly in the metaphors of Francis Bacon, to be vexed,
interrogated, penetrated, bound into service, and tormented
before finally ending up as an inanimate mechanism.[18]

Against all this backdrop, *cunt* became a quintessentially
"dirty word." *Dirt* itself has multiple meanings. *Dirt* is the very
ground under our feet, *earth* or soil (*OED*). It also functions meta-
phorically, in an everyday pornographic culture,[19] to signify the
supposedly inferior, unintelligent, flawed, contaminating, and
spiritually and intellectually insignificant realms of materiality,
sexuality, carnality, and animality.

A hierarchical split stigmatizes all that those who claim
themselves to be "clean" name to be "dirty." The whole sorry setup
denies that humans owe our lives to an origination and a culmina-
tion in darkness, that we are carnal beings born of sex and women,
that we are animals made of Earth, and that sexual pleasure as
well as processes of aging, dying, and disintegrating are all ways
of knowledge. The reality of our "animality," political scientist
and disability theorist Ruth O'Brien recognizes, *is* our humanity,
including our inherent flaws, frailty, and need for care: "There is
no such thing as a normal healthy mind and body . . . nor should
a normal mind and body constitute an ideal to which members
of society should aspire. Creating such an ideal turns human

needs into weaknesses . . . rather than one's humanness."[20] All such alleged weaknesses then can be targeted for "cleansings" of one kind or another by those who fantasize themselves as pure, perfect, beautiful, the norm, and the strong. Vanessa Watts (Bear Clan from Mohawk and Anishnaabe Nations) reminds those human beings who have forgotten that we all "are made from the land, our flesh is literally an extension of soil." Being "made of the stuff of soil and spirit,"[21] people carry profound obligations to land.

Ecofeminists, ecowomanists, and related thinkers, including Susan Griffin, Delores Williams, Gloria Anzaldúa, Vandana Shiva, Val Plumwood, Maria Mies, Greta Gaard, Barbara Mor, and Donna Haraway have denounced the Western hierarchical dualism that forms invidious oppositions between dirty and clean, nature and culture, soil and spirit, black and white, earth and sky, female and male.[22] Mor directs attention to a wholly other understanding that manifests in an African "original Black Goddess," the Dahomey "Mawu-Lisa, imaged as a serpent; Mawu-Lisa was both female and male, self-fertilizing, seen as the earth and the rainbow."[23]

Historian Sterling Stuckey observes that it was the insistence of "African religion . . . upon the unity of seeming opposites" that caused Europeans to deem these belief systems "savage,"[24] an accusation the invaders then used as a justification for slavery and colonization. Atrocities are regularly legitimated in this way. Genocidal killers perform "ethnic cleansing." Sex killers target those women whom they deem born "whores" and "dirty sluts."[25] Ecocide is masked as the light of civilization triumphing over dark savagery. The Western logic regarding the dark and the dirty is not only invidious, it's nonsensical. Where would we be without them? Nothing could grow, sleep, or dream. We'd all burn up.

These negations also sever sex from the sacred. Multiple feminist, lesbian, disability rights, and queer and funk thinkers blast the normative "erotophobia"[26] that deems sex "a danger-ous, destructive, sinful, negative force"[27]—except for when it is heterosexual, marital, monogamous, and not too pleasurable. Erotophobia makes of sex something "lower" and opposed to "higher" mind or spirit. Sex is said to be even filthier when located in the bodies of people of color and/or outside of heteropatriarchal norms. Theologian elias farajajé-jones finds "the roots of religious erotophobia" in prevailing "notions of a desexed God, a basically bodiless God, without any notion of Goddess."[28] This theology then "intersects with White supremacy in the investment of peo-ples of color as the exotic/erotic other—people who are . . . sexu-ally dangerous and in need of domination."[29]

Scratch the surface of most of our culture's "dirty" or at least highly insulting words—*cunt, cocksucker, cumdumpster, cunt-face, horseface, cow, heifer, chick, dog, pig, ape, douchebag, lame, retard, son of a bitch, bitch ass, ass-licker, fat-ass, shit, bitch, ass-hole, freak*—and you will sniff out mutually reinforcing erotopho-bia, homophobia, misogyny, matriphobia, transphobia, ableism, racism, ethnocentrism, speciesism, and biophobia.[30] Genocidal German Nazis, for example, explicitly referred to Jews as insect-like and inherently dirty. To oppress another human, you define and treat them like "dirt," a "freak (of nature)," an "animal," and then you subject them to some form of moral, ethnic, or "socially hygienic" (eugenic) cleansing or purification.

Novelist Louise Erdrich (Ojibwe) offers this thought: "After all, we live on earth. We are created of the earth. The Ojibwe word for the human vagina is derived from the word for earth. A profound connection, don't you think?"[31] In a culture that loves women and respects the life force-source, this connection leads to respect. In the misogynist *Man's* world, though, the Earth womb-tomb is envied and feared—put down, ripped up, ripped off, and raped. Earth still is recognized (more or less openly) as a primal cunt, but this now means that Earth gets *fucked* in the all-out matricidal and deicidal assault that is the Anthropocene.

The writer John Updike, assuming the position of spokesman for men, once proclaimed: "We want to fuck what we fear."[32] Such "fucking," as critiqued indelibly by Andrea Dworkin,[33] allows men to experience their masculinity as a violent victory over the feminine, establishing total control and getting sexual satisfaction from so doing. *Fuck* and related slang terms like *screw, plow, nail, bang, slice, split, hit, drill*, and *beat up the pussy* reveal the heteropatriarchal fusion of sex and violence as well as the sexualized and gendered dynamic that informs domination, possession, and violation from the personal to the planetary. The *fucker* traditionally is "the man," and the *fuckee* is "the girl," "the pussy," "the bitch," or in this case "the mother" (Nature-Earth) against whom all manner of sexual violation is aimed, right up to nuclear annihilation.[34]

Right now, angry women and allies are saying "Me Too" and "Time's Up," as is Mother Nature-Earth.[35] Public intellectual and activist Tina Ngata (Ngāti Porou) makes precisely this connection: "Because if time is up on the exploitation of women, then it must surely also be up on the exploitation of our region."[36] Ngata is referring specifically to the Pacific islands, but the sentiment holds true for the planet.

Upending *The Man's* worldview entails a reclamation of "the erotic," named by lesbian and Black feminist poet-theorist Audre Lorde as the source for the "power" that fuels resistance, inner cohesion, creativity, and positive "change."[37] To reclaim the erotic is to get to the *nitty gritty*, to the "heart of the matter" (*OED*), where one encounters—within and without—the funky force field of Nature-Earth. In this way, it's possible to put the *dirty* back where it belongs—into the *mind*. A *dirty mind* perceives the inter-connectedness of all, including the dark in the light, the female in the male, the sensual in the intellectual, the obscene in the sacred, the cunt in the cock, the down in the up, the animal in the human, the sexual in the spiritual—and vice versa.

THE EARTH "MUTHA'"

In their smart and scathing book, *Articulate While Black*, scholars Samy H. Alim and Geneva Smitherman note the intense popularity in "African American Language" of *motherfucker*. They also recount the Black folk theory regarding its origins:

> Black folk etymological theory posits that *motherfucker* was a term developed by the children of enslaved Africans in the United States . . . as the best way to refer to White slavemasters, who enacted a particularly savage form of physical and sexual abuse on the bodies and spirits of Black women . . . It was used to describe White men who raped your mother in order to break the Black family down, physically and psychologically, and as a means to avoid calling your slavemaster your "father." Muthafucka, as the theory goes, then became the worst thing you could call somebody.[38]

The chattel slave system ensured that both an enslaved mother and her children were the master's legal property. Christina Sharpe explains that this was mandated by the rule of *"partus sequitur ventrem* (that which is brought forth follows the womb), in which the Black child inherits the non/status, the non/being of the mother."[39] The master, then, is the *motherfucker,* the model bad father, and his heinous motherfucking epitomizes an ongoing pattern of sexual taking, spirit breaking, profit making, and toxic wasting.

This "worst" use of *motherfucker* has a great deal to say about what is going on in the Anthropocene, with explicit recognition of this found in literary, visual, and popular culture.[40] *Motherfucker* as the proper name for *The Man* in all of his depredations appears in June Jordan's blazing 1972 poem "Getting Down to Get Over," as she demands to know: "What does Mothafuckin mean?/ WHO'S THE MOTHAFUCKA/ FUCKED MY MOMMA . . . the first primordial/ the paradig/digmatic/ dogmatistic mothafucka who/ is he?"[41] In 2003, Hip Hop musician André 3000 connected the word directly to what's going on in the Anthropocene: "And when I say motherfucker I do mean motherfucker/Because Mother Earth is dying and we continue to fuck her to death."[42] Both artists use what's called *strong language*.[43] But isn't such strength imperative? Listeners, knowing and feeling strongly, might do something about ending the gendered, classed, sexualized, and racialized violence characterizing *The Man's* world. In so doing, we get angry, speak out, curse out, demanding a stop to all this.

Motherfucker, like all true words, is alive and evolving. As time passed, the word turned, showing another face and revealing new meanings. Black English characteristically turns on ambiguity, innovation, and inversion, extending "the meaning of words on multiple levels."[44] The word, again, now also means "a formidable, strong, or otherwise admirable person (chiefly in African American use) . . . A large, impressive, or outstanding example of something" (*OED*).

In a 1977 album, *Mutha's Nature*, James Brown was perhaps the first to explicitly connect Nature-Earth with the "Mutha'," lamenting what he calls the "war" being waged against nature.[45] A few decades later, the New Zealand band House of Downtown issued the album *Mutha Funkin Earth*, naming Earth as a "paradise."[46] I had not known about either of these when I started this project. Although I previously had considered the ecological significance of *motherfucker*,[47] the recognition of the Earth as *"Mutha'"* first came to me when reading the formidable funk theorist L. H. Stallings's *Mutha' Is Half a Word: Intersections of Folklore, Vernacular, Myth, and Queerness in Black Female Culture*. Stallings brings together trickster traditions, which include tendencies to confound oppositional dualism, to combine "orality and sexuality," and to enjoy the word *muthafucka*. These come to bear in the empowering verbal play of "trickster-troping" Black women as they "continuously strive for radical Black female sexual subjectivities that are just right for them."[48]

In this way, listen to funk singer, songwriter, and actor Janelle Monáe: "Being a queer black woman in America, someone who has been in relationships with both men and women—I consider myself to be a free-ass motherfucker."[49]

Stallings claims *muthafucka* as a "sacredly profane word, and one that . . . can discombobulate people and their discourses of nation, gender, sexuality, race, and class."[50] I respectfully follow Stallings's lead and hope that calling Nature-Earth *"Mutha'"* causes continuing discombobulation, especially as this force-source emerges ever more into stories, artworks, theories, practices, oracular dreams, and visions as the absolute exemplar of the "free-assed," self-possessed subject.

The twists and turns of *motherfucker* derive from what Wynter deems "the incredible inventiveness of black culture" understood within "the imperative task of transformation."[51] This necessary concept of the "Mutha'" is synonymous with Mother Nature-Earth, but spoken and heard in a manner that refuses the hateful stereotypes. Astoundingly, the word both reveals how the Earth has come to its present form of crisis in the Anthropocene and, at the same time, bespeaks a way out. Earth "Mutha'" values speak not only to environmental justice, but to all contemporary social justice movements grounded in spirited recognition of the being-ness, connectedness and integrity of others (human and other than human), self-love and self-generation,[52] mutuality, reciprocity, respect, interdependence, and community.[53]

For example, Patrisse Cullors, a co-founder of the Movement for Black Lives, affirms that climate change is central to that movement because climate change *is* a race issue, caused by a racist system and with worse effects on people of color.[54] At the same time, Cullors declares this movement to be necessarily grounded in "spirit," but not *spirit* as something dissociated from and superior to matter: "I don't believe spirit is this thing that lives outside of us dictating our lives, but rather our ability to be deeply connected to something that is bigger than us. I think that is what makes our work powerful."[55] This is down-to-Earth spirit, the energy that connects all terrestrial life, allowing relationships. That interconnectedness allows humans to "listen in,"[56] to know and hear the speech of all beings and to give heed to the warning now being issued to protect and serve the force-source that enables all of us to be. Cultures made in the image of *Man* have falsely sundered spirit and matter. The "spirit" or "something bigger" that Cullors invokes is the *ground* of being, the shared basis of all who constitute Mother Nature-Earth.

A question remains as to the proper pronoun for this Mother. Official grammars set up a dualism of "he-him" in opposition to "she-her," with the implicit understanding that to be a real *he* means that *he* overwhelms *her*. Welcome alternatives originated by trans and non-binary gender activists now generate at the speed of the light of the rainbow.[57] Many use the pronoun *they* in the singular to refute the oppositional dualism and open up understandings of the plurality of being. Pronouns like *fae, faers*, and *faer*[58] acknowledge not only non-binary realities, but also kinship with elemental beings (in this case the faerie folk). Reflecting all terrestrial life, the "Mutha'" is female, femme, non-binary, multiple, and a volatile essence.[59] As such, and as the feeling takes me, I use pronouns *she, they*, and sometimes ones made up on the spot—like *S/H/**!*—to hail this force-source.

In 1976, Lesbian Feminist artist, thinker, and activist Liza Cowan created a button reading "Mother Nature is a lesbian" to express her resistance to heteropatriarchy.[60] The celebrated drag queen, the Lady Bunny, gives a different spin, expressing gratitude for the beautiful weather at the annual (from the 1980s to the 1990s) Wigstock Festival in Greenwich Village and exclaiming, "I think Mother Nature must be a drag queen."[61] Ecosexual theorists and activists Elizabeth M. Stephens and Annie M. Sprinkle understand Earth as *Lover*: "The earth is our lover. We are madly, passionately and fiercely in love, and we are grateful for this relationship every day."[62] For some, Nature-Earth is best figured as clearly non-human, as owl, tree, glacier, and/or enveloping cosmos. These understandings are all different, carrying creative tension as comprehension grows. Underlying all of them, though, is a commonality,[63] a respect for the being, integrity, and self-driving nature of the force-source. In that necessary context, "Mutha'" Nature is, perhaps, who they appear to you to be and whom you find in yourself.

A quintessential conjuring of this force-source marks Amir Khadar's artwork honoring the 2017 Trans Day of Resilience, an annual event on November 20 mourning transphobic violence—which particularly targets Black and Latinx femmes—and affirming survival into a more just future.[64] (See Figure 2.1.) The collaged image shows a Black femme walking on a swirling sea, pouring libations from pitchers held in each hand. Flowers bloom and rolling hills arise on either side. This being is bejeweled and draped with the shining stars of the black night sky, attracting others into their gravitational field. A flame-shaped aura of pink and purple (colors associated with the Yoruban Oya, deity of change, weather, wind, and fire) emanates from their upper body. A banner reads: "& Like Any Goddexx You Are Scorned & Become the Fire Anyway." Khadar's artwork evokes the elemental context who is the convergence and emergence of all being. This "Mutha'" is energy and matter, essence and change, dark and light, solution and mystery. *She, They, Fae, S/H**!* is the power capable of undoing *Man*.

CALLING THE "MUTHA'"

I first realized the need to call Mother Nature-Earth by name some decades ago, when listening to Marilou Awiakta (Cherokee) speak her admonitory poem "When Earth Becomes an 'It.'"[65] Awiakta makes it clear that Mother Earth, in the face of disrespect, can just forget about us, and go away. This autonomous[66] force-source thinks, decides, gives, takes, comes and also can go. Human beings have long named Nature-Earth "Mother" not only to recognize our dependence on and connection to Nature-Earth, but also, perhaps, to remind Nature-Earth of their relationship

FIGURE 2.1 Amir Khadar, *Trans Day of Resilience Poster* (2017). Digital Illustration & Collage.

to us. Think about it—the worst thing that could happen to us would be if the Mother ceased to think about, care about, or even be much aware of us. Actually, this might be exactly "what's going on" in the Anthropocene.

The Earth "Mutha'" story now is being heard and told by dreamers, activists, tale-tellers, shapeshifters, poets, teachers, bloggers, rappers, artists, musicians, songwriters, singers, jokesters, street talkers, scholars, seers, listeners, and visionaries of various stripes. It assuredly is not a kind of Old Testament punitive divine retribution tale, nor is it a Hollywood rape revenge narrative.[67] It is, instead, a *Call Your "Mutha'"* story, to call as in to cry out—in multiple ways and voices, individual and collective—to tell the truth about what is going on, to attract attention, to ask forgiveness, to implore help, to pay what we owe, to respect and fulfill our spiritual/material obligations to that "something bigger" from whom we all came and to whom we necessarily return.

Awiakta instructs that to thwart calamity we need to "call" Mother Earth "by her name." She is not advising the simple utterance of a name. To *call by name* is to *invoke*. This is ritual speech designed to appeal to and conjure the force-source, known by many names, and to mention only a few: *Maamé,*[68] *"Mother Love, Ma Dear, Earth Mother,"*[69] *"Big Mama Nature,"*[70] *Madre Tierra,*[71] *Nunamshua, The "Indweller of Earth,"*[72] *Medusa,*[73] *Hine,*[74] *Cihuateotle,*[75] *Atabey,*[76] *Sila,*[77] *Oya,*[78] *Kali,*[79] *Amalur,*[80] *Changing Woman,*[81] *Nüwa,*[82] *Demeter,*[83] *Ala,*[84] *Muzzu-Kummik-Quae,*[85] *Bhumi,*[86] *Pachamama, Papahānaumoku,*[87] *Unci Maka/Grandmother Earth,*[88] *Tonantzin/ Guadalupe,*[89] *Inang Kalikasan,*[90] *"Mutha'" Nature, "Mutha'" Earth.*

PRELUDE TO A CURSE

Multiple wisdom traditions speak of magic words, including a time when all words were potentially so.[1] A few words still are. Sometimes only one is needed if you are wise. *Motherfucker* might well be one of these.

The following section is inspired by that word's extraordinary capacity as a "curse word." This is one "invoking a deity" while "damning or punishing someone."[2] The recipient of the imprecation here is The Age of *Man* in all its manifestations. The deity called on is Mother Nature-Earth, our planet/ourselves.

The first chapter calls out the heavenly patriarchal God, the mask of *Man*, naming his intent to conquer nature for what it is— attempted deicide in the form of rape-murder against his earthy rival, Mother Nature-Earth. The second addresses the mother-fucking paradigm[3] enacted in the founding of the United States and the paving of the way to the Anthropocene. The third considers monochromatic *Man's* whitening of the world, epitomizing his intent to kill off, copy, and replace nature.

Considering current dire conditions, critique is not enough. A curse most fully names what's going on, also gathering force to go another way.

THE CURSE

"Late Old English *curs*, of unknown origin . . . generally the opposite of *to bless* . . . to utter against." (*OED*)

THE GODS WE WORSHIP

THE EXTRA-KEEN PERCEIVER RACHEL CARSON, speaking in 1962, wisely cautioned against arrogant humans trying to impose their "will on Nature." She insisted, instead, on the need to be "quiet and listen to what she has to tell us."[1] The following year, in a preface to Ruth Harrison's *Animal Machines*, an exposé of the atrocious practices of industrial farming, Carson further warned: "The modern world worships the gods of speed and quantity, and of the quick and easy profit, and out of this idolatry monstrous evils have arisen."[2]

This worship of disastrous "gods" dominates this current time period that earth scientists call the *Great Acceleration*.[3] This refers to the runaway rate of human activity affecting earth beings and systems since 1950, resulting in a tremendous increase in wealth disparity among and within nations, an exponentially growing world human population, a disability-producing "pace of life,"[4] ever more lethal weaponry, and the production of more and more consumer stuff, all of which lead to increased carbon emissions, meat consumption, species extinctions, the pollution of air and water, and the obliteration of forests and wetlands. There is a corresponding uptick in human-generated wastes, material and psychical—harmful chemicals, radioactive particles,[5] and, it seems, augmented stress, rage, depression, and loneliness.

The most worshipful of Anthropocene commentators—many of whom are White, highly educated, and far removed from the damages wrought by their reverence for monstrous gods—do recognize some of these problems. However, they evade responsibility,

Call Your "Mutha'". Jane Caputi, Oxford University Press (2020). © Oxford University Press.
DOI: 10.1093/oso/9780190902704.001.0001

masking structural inequities and their own culpability by referring to a universal "we" of "humans" doing the damage. They cal-
culate the torments and toxicities as regrettable but unintended
side effects of civilization and progress. They predict miraculous
technological fixes just on the horizon, devised by what they call
the uniquely human capacity for ingenuity and invention. They
claim innocence because they did not really mean it, did not know
what they were doing, or were too immature to be responsible. At
the same time, more or less openly, they praise "humanity" for
being just so almighty.

Best-selling US science writer Diana Ackerman, in *The
Human Age: The World Shaped by Us*, euphemizes ecological
crimes like taking all the food, stealing water, poisoning the world
and then dumping toxic wastes on other, poorer peoples, chalking
these activities up to just a phase of "human" childhood: "Without
really meaning to, we have nearly emptied the world's pantry, left
all the taps running, torn the furniture, strewn our old toys where
they're becoming a menace, polluted and spilled and generally
messed up our planetary home."[6]

Another apologist, astrobiologist David Grinspoon in *The Earth
in Human Hands*, relies on similar exculpatory metaphors: "We
suddenly find ourselves sort of running a planet—a role we never
anticipated or sought."[7] As he tells it, "humans" just "obliviously
stumble[d] into ecological and climate danger," and just "inad-
vertently messing with Earth." However, he assures readers, such
"humans" are essentially blameless, not criminals but more like
an "orphaned baby," "an infant staring at its hands." If that doesn't
convince, Grinspoon tries a teenage angst angle: "We're like an
awkward, naïve, and reckless adolescent . . . [with] little awareness
of our limits, and we love to watch things blow up. . . We're not
criminals committing premeditated planetary desecration. We're
more like juvenile delinquents in need of rehabilitation."[8]

All these messy children and needy adolescents are implicitly
White, male, and affluent. If not, they would be taken from their
parents and put into child services; they would be locked up, committed, even shot by the police. Moreover, the excuses Ackerman

and Grinspoon proffer are the same ones claimed by rapists, domestic batterers, harassers and their apologists—just the "'nice guy'" who accidentally "messes with" (rapes) someone, the abuser (of any age) whose bright future shouldn't be clouded because of his "mistake," the loving batterer, who "didn't really mean it" and won't ever do it again—until he does.[9] All this makes the culprit's supposedly innocent intentionality far more important than any of his disastrous impacts, while also foregrounding his perspectives over those of the victimized.

These non-apologies for the Anthropocene's damages are accompanied by outright self-congratulation. Diane Ackerman can't help but be impressed with the ways that "the human race" has so stunningly become "the single dominant force of change on the planet."[10] Reciting a long list of human accomplishments, she concludes: "We've turned the landscape into another form of architecture; we've made the planet our sandbox."[11] She does express supercilious solicitousness for the source of life:

> Nature is still our mother, but she's grown older and less independent. We've grown more self-reliant . . . We may not worship Mother Nature, but we love and respect her, are fascinated by her secrets, worry about alienating her, fear her harshest moods, cannot survive without her. As we're becoming acutely aware of just how vulnerable she truly is . . . we're trying to grow into the role of loving caregivers.[12]

Even while claiming love, respect, and awareness of dependence, Ackerman caricatures Mother Nature as a kind of failing granny on the way out. Meanwhile, David Grinspoon sounds like the White man declaring a new burden for himself: "We will either have to play god or allow species to go extinct (which, I suppose, may still be a form of playing god)."[13]

An instructional film about the Anthropocene produced by the Commonwealth Scientific and Industrial Research Organization (CSIRO) for a 2012 UN conference tells much the same distorted story. A soft-voiced woman narrates, intoning the achievements of "humans": "In a single lifetime we have grown into a phenomenal global force. We move more sediment and rock annually than all natural processes such as erosion and rivers . . . We are altering Earth's natural cycles. We have entered the Anthropocene."[14] Scholar Gabrielle Hecht scorns such *Man*-aggrandizing soundbites and brings us back down to the environmentally unjust realities: "Who actually moved the rock? How did this movement affect the people and ecosystems around the mines, not just at the time of extraction but decades later?"[15]

Hecht's challenges long have been voiced by a global environmental justice movement, whose principles were affirmed at a 1991 multinational summit devoted to People of Color's Environmental Leadership. These include the need "to build a national and international movement of all peoples of color to fight the destruction and taking of our lands and communities." The statement of principles refers to the environmental threats posed by racism, as well as to the spiritual significance of these threats. It affirms the "ecological unity and the interdependence of all species, and the right to be free from ecological destruction."[16] It also calls for human societies to "re-establish our spiritual interdependence to the sacredness of our Mother Earth." *Man*, however, officially denies that sacredness as part of his ultimate stratagem to replace Mother Nature-Earth.

In the Anthropocene, the top tier settle into luxury high-rises looking down on their "sandbox" world, able to profit without actually getting their hands dirty. They set a pace that cannot be sustained and hoard all the goods and wealth along with liberty, justice, security, safety and health that allow people to thrive.[17] Their riches pile up to the extent that others are impoverished; they make waste and dump it in someone else's neighborhood. In their minds, it seems, they are smart, powerful and godlike enough to enact change on a planetary scale, yet too innocent and bumbling to be held responsible for the damages. In worshipping those disastrous gods so insightfully named by Carson, they, basically, are worshipping themselves.

THE "GUIDING VISION"

Worship, Peggy Beck and Anna Lee Walters (Pawnee and Otoe-Missouria) explain, always involves a "guiding vision" based in what a people hold sacred, worthy of reverence and respect.[18] People maintain, strengthen, and renew this vision via personal and public devotions—employing images, words, votive offerings, and "appropriate acts, rites, or ceremonies."[19]

Such guiding visions have considerable world-building power. Paul Crutzen and Christian Schwägerl refer without much ado to the vision that has ushered in the Anthropocene: "A long-held religious and philosophical idea—humans as the masters of planet Earth—has turned into a stark reality. . . . It's no longer us against "Nature." Instead, it's we who decide what nature is and what it will be."[20] The Anthropocene, supposedly so new and modern, is not at all.

Man's sacred mandate to dominate all defined as "nature" appears in the biblical book of Genesis, where God grants Adam and Eve dominion over the earth, with Adam retaining the upper hand on Eve. It appears in philosophy in Rene Descartes who proclaimed *Man's* goal "to render ourselves master and possessor of Nature"[21] and continues into Slavoj Žižek's contemporary quip, "If there is one good thing about capitalism it is that under it, Mother Earth no longer exists."[22]

European culture was not always so cavalier regarding its relationship to Mother Nature-Earth. Carolyn Merchant explains that an "organic worldview" prevailed prior to the Scientific Revolution in which Earth was understood to be a "living organism," a "mother," albeit in the context of sexist stereotypes—on the one hand, passive, chaste, kindly, nurturing, and good and, on the other, autonomous, unruly, disorderly, a "common harlot,"[23] the ultimate bad woman. The organic worldview, though, did proffer Earth at least some respect and attendant constraints against exploitation.

In the 17th century things changed. Francis Bacon, Merchant writes, "In bold sexual imagery, outlined the key features of the modem experimental method—constraint of nature in the laboratory, dissection by hand and mind, and the penetration of nature's hidden secrets—language still used today in praising a scientist's 'hard facts,' 'penetrating mind,' or 'seminal' arguments. The constraints against mining the earth were subtly turned into sanctions for exploiting and 'raping' nature for human good."[24] Enthusiasm grew for bringing an unruly Mother Nature to heel. Meanwhile, passive Mother Earth was no longer off limits for exploration and exploitation; her "womb" could now be penetrated, her "secrets" unlocked.

Merchant concludes that this "new image of nature as a female to be controlled and dissected through experiment legitimated the exploitation of natural resources."[25] The organic worldview gave way to a "mechanical worldview," ordaining the "death of nature."[26] This defined Earth as an inert object void of agency, purpose, will, or spirit, something to be dissected and reduced to parts that could be manipulated, rearranged, and even made anew. Any problems with this approach, the mechanical worldview promised, could be remedied by a "technological fix." This discourse continues today as nuclear physicists and their chroniclers speak of investigating the "most intimate properties of matter," tearing away "the veils," and forcing open "a resistant atom."[27] Synthetic biologists and biochemists march along that same path, promising to "unlock" the deepest of "nature's secrets"[28] held in the genetic code.

The earlier European "organic worldview" acknowledged spirit and sentience in the natural world: "Every tree, every spring, every stream, every hill had its own *genius loci*, its guardian spirit,"[29] historian Lynn White explains. But a subsequent denial of these beings, a veritable dis-spiriting[30] of Nature, is part of a complex of theological and philosophical tenets that have enabled ecological destruction—allowing some humans to torture other animals, slash forests, contaminate rivers, and push greenhouse gases into the air with no moral judgment attached. This creed of human exceptionalism denies the necessity, scholar Vanessa Watts writes, of the "treaty making and historical agreements that human beings held with the animal world, the sky world, the spirit world, etc."[31] These agreements assure, as much as can be expected in an inherently unpredictable world, ethical behavior among all parties as well as food, habitat, well-being, and continuance for all.

Intrinsic to Christian theology was a condemnation of nature spirits, or *daemons*, as evil *demons*. The apostle St. Paul (who died in 64 CE) and the influential church father St. Augustine of Hippo (354–430) wrote vehemently against them,[32] teachings ultimately leading to mass razing of sacred groves and abandonment of rites of worship at sacred springs, particularly during the Witch Burning period, 1450–1750. White writes: "By destroying pagan animism, Christianity made it possible to exploit nature in a mood of indifference to the feelings of natural objects."[33] Nowadays, corporations and governments routinely contaminate rivers, decapitate mountains, lay waste to species, and mow down forests. Though not openly acknowledged, these acts of ecocide constitute sacrificial offerings to those Anthropocene gods of "speed and quantity, and of the quick and easy profit" (Carson), gods of *more, me, mine, faster, bigger, newer,* and *now.*

Contemporary spokesmen for the Anthropocene like Crutzen and Schwägerl go on to suggest an update to the traditional guiding vision: "We should shift our mission from crusade to management, so we can steer nature's course symbiotically instead of enslaving the formerly natural world."[34] The older plantation-speak of "enslaving" is thus replaced with the new corporate-speak of "steering" and "managing," but the slave-master model remains unchanged. At the same time, Crutzen and Schwägerl envision the final elimination of autonomous living Mother Nature-Earth as both a good and a given, evidenced in their easy reference to "the formerly natural world."

This phrase suggests *deicide* if we keep in mind Tamra Andrews's reminder that, "Almost every ancient culture deified the Earth [and] . . . deified the earth as female."[35] The guiding vision of the Anthropocene rests upon the theological notion that the heavenly father God made humans (again, really *Man*) in his image, stripping divinity from trees, corn, waters, stones, buffalo, fish, and rivers, while also seeking to destroy the spirit and will of all who reverence those beings.

Novelist N. Scott Momaday (Kiowa) writes that "The Buffalo was the animal representation of the sun, the essential and sacrificial victim of the Sun Dance. When the wild herds were destroyed, so too was the will of the Kiowa people." Momaday tells of his grandmother, who at the age of ten attended the last Sun Dance. "They could find no buffalo; they had to hang an old hide from the sacred tree . . . Forbidden without cause the essential act of their faith, having seen the wild herds slaughtered and left to rot upon the ground, the Kiowas backed away forever from the medicine tree. That was July 20, 1890, at the great bend of the Washita. My grandmother was there. Without bitterness, and for as long as she lived, she bore a vision of deicide."[36]

As *Man* kills off those rival deities, he also fancies himself the new god. Historian Yuval Noah Harari, in his bestselling *Homo Deus*, writes that "with regard to other animals, human have long since become gods."[37] Under this sorry framework, to be godlike is to be able to wipe out the vast majority of wild animals on the planet, while also torturing billions of others for profit. This equation of "god" with absolute power over everybody else is in keeping with traditional patriarchal theology that venerates omnipotence, omniscience, omnipresence, immortality, and perfection. But these attributes don't represent any *de facto* understandings of divinity. Indeed, for many of us they might instead smack of psychopathy. The reigning model of the male god is modeled on *The Man*, the lord and master. In the Anthropocene any pretense of a distinction between ruling men and the god they have made in their own image is dropped. *Man* now openly declares himself *God* and demands devotion.

GOD PLAY

My first book, *The Age of Sex Crime*, concerns the spiritual and political meanings of serial sex killers—notorious men, like Jack the Ripper.[38] These killers very often are mythicized by their chroniclers as outlaw heroes, preternatural entities, geniuses, and even immortal gods. Some killers describe themselves in much the same ways. Since then, I have been alert to deification discourses surrounding violent and misogynist men, which, as the following examples show, are all too common:

> "I was making life and death decisions . . . playing God in their lives." —Edmund Kemper, the "Coed Killer"[39]

> "I viewed myself as God basically and she was my slave." —Josh Brown, National Football League player describing his mindset when abusing his wife[40]

> "Ich bin Gott" – I am God." —Eric Harris, Columbine High School killer, writing in a friend's yearbook[41]

> "Call Me God." —John Allen Muhammed, "The DC Sniper," who committed mass murder as part of a plan to kill his estranged wife[42]

> "Magnificent, glorious, supreme, eminent ... Divine! I am the closest thing there is to a living god." —Elliot Rodger, misogynist and mass murderer[43]

> "That godlike feeling you had was in the field. It was like I was God. I could take a life. I could screw a woman." —Anonymous Vietnam veteran[44]

Some years later, reading up on the Anthropocene, I found similar statements, this time from highly educated and respected men:

> "Men have become like gods. Isn't it about time that we understood our divinity? Science offers us total mastery over our environment and over our destiny." —Anthropologist Edmund Leach, author of *A Runaway World?* (1968)[45]

> "We are as gods and have to get good at it." —Stewart Brand, eco-modernist[46]

> "We humans are the God species, both the creators and destroyers of life on this planet. As we enter a new geological era—the

Anthropocene—our collective power now overwhelms and dominates the major forces of nature." —Science writer Mark Lynas, author of *The God Species: Saving the Planet in the Age of Humans* (2011)[47]

"So as we evolve, we become closer to God. . . . There is beauty and love and creativity and intelligence in the world—it all comes from the neocortex. So we're going to expand the brain's neocortex and become more godlike." —Ray Kurzweil, inventor and author[48]

The upgrading of humans into gods may follow any of three paths: biological engineering, cyborg engineering and the engineering of non-organic beings." —Yuval Harari, historian and author of *Homo Deus* (2015)[49]

Man's self-deification can be more subtle. David Grinspoon's title, *Earth in Human Hands,* recalls the traditional spiritual where "God" has "the whole world in His hands."[50] A similar conceit appears in any number of pop-culture images, including the illustration on a National Public Radio webpage devoted to the Anthropocene, showing Earth from space, filled with tall skyscrapers and launching missiles. Underneath the planet is a giant White and stereotypically masculine-type hand.[51]

The cover of *Earth in Human Hands* displays its title in big, bold, black capital letters. Behind that, if you look closely, is a faint image of the whole Earth. Grinspoon muses that such photos of Earth are shifting humanity toward an identification with the planet and a sense of profound connection. These photos may indeed have inspired some people in that way; for others, though, the same images seem instead to inspire a desire to lord it over the planet, to openly identify with the God who reportedly made the Earth rather than with the Earth. Much like Grinspoon's reckless "adolescents," this God also wants to "blow things up."

This final blasting of Earth by God is prophesied in the Gospel of St. Peter: "The day of the Lord comes like a thief, and then the heavens will pass away with a loud noise, and the elements will be dissolved with fire, and the earth and the works that are upon it will be burned up." Mary Daly quotes this passage, deeming it to be a "self-fulling prophecy and manifesto of necrophilic faith."[52] She means this as a life-loathing patriarchal belief system that desires fixity and objectification—a controlled woman, animal being, or landscape. This life-hating faith also worships man-made destruction, understood as sexually gratifying and sacred.[53] Nuclear weaponry, now capable of burning up the Earth, makes this all too possible.

This God, after all, is said to have made Earth so it stands to reason that he also can take Earth out. A postcard shows an objectified Earth viewed from space and proclaims "Father God created Mother Earth."[54] Centuries earlier, a medieval artist portrayed this same God and a similarly positioned Earth in a well-known frontispiece for a late 13th-century bible, "God as Architect/Builder/Geometer/Craftsman"[55] (Figure 3.1). This space-based God holds a tiny Earth in one hand. With the other hand, he "inscribes with a compass the newly-created circle of the world." Inside that circle, art historian Mary Garrard explains, is "*hyle*, the original chaos or matter out of which God produced architectural order." This theology entails "the 'imposition of designed form upon lifeless matter by a knowing creator.' " Crucially, Garrard reminds us that this scenario reverses previous religious understanding of *Nature* as the "engenderer."[56]

In the 16th century, a resonantly similar objectified image of Earth appeared in a frontispiece for a Spanish conquistador manual.[57] On the right appears a smug ship's captain. With one hand, he grips a sword. With the other, he places a compass on the surface of a globe. The inscription reads "*A la Espada y el Compas, Mas y Mas y Mas y Mas*" (With the sword and the compass, more and more and more and more; Figure 3.2). A 1493 bull issued by Pope Alexander VI gave sacred sanction to this imperial conquest. It read in part: "We trust in Him from whom empires and governments and all good things proceed, that, should you, with the Lord's guidance, pursue this holy and praiseworthy undertaking."[58]

FIGURE 3.1 *God as Architect/Builder/Geometer/Craftsman.* From *Art in the Christian Tradition*, a project of the Vanderbilt Divinity Library, Nashville, Tennessee, http://diglib.library.vanderbilt.edu/act-imagelink.pl?RC=55539. Retrieved June 27, 2019.

A la Espada y el compas
Mas y mas y mas y mas

FIGURE 3.2 Frontispiece for a Spanish conquistador manual (1599). From Bernardo de Vargas Machuca, *Milicia y descripcion de las Indias*, Madrid (1599). Courtesy of John Carter Brown Library at Brown University.

The accelerating expansion of that "holy" undertaking culminates in the Anthropocene, often also illustrated with the image of a conquered and carved up Earth. A cover of *The Economist* (May 28–June 3, 2011) announces: "Welcome to the Anthropocene."[59] Viewers see a drawing of the Earth from space. The planet is mostly whole, but if you look closely, some small parts of it are just gone. The Earth's surface, too, is radically different, now consisting of riveted sections of metal. Several of these are removed and lattice-like structures show Earth's previously hidden depths. There is no need to include a male figure representing either God or *conquistador*. The word *Anthropocene* says it all. "Anthropos" himself is now Architect/Builder/Geometer/Craftsman, able to create and destroy the planet. *Man* is now God. *Mas y Mas y Mas y Mas.*

TOWERS OF (PHALLIC) POWER

Kevin Kelly, the founder of *Wired* magazine, a publication devoted to the culture of new technologies, prophesies a coming "technium . . . a self-reinforcing system of creation" in which tools and machines become conscious, reproduce themselves, attain autonomy, and evolve beyond humanity.[60] He predicts that "in the not-too-distant future the magnificence of certain patches of the technium will rival the splendors of the natural world. We will rhapsodize about this or that technology's charms and marvel at its subtlety. We will travel to it with children in tow to sit in silence beneath its towers."[61]

Kelly predicts future outright worship of lordly techno-towers, but, even now, towering skyscrapers are a common signifier of the Anthropocene. One reason for this is because skyscrapers communicate so clearly the ideology of human exceptionalism. Writing in 2017 in the *Chronicle of Higher Education*, journalist Tom Bartlett enthuses: "Humans design skyscrapers, compose operas, and write sonnets. That makes us different than—indeed, superior to—the rest of the animals on the planet."[62] Is this really so? Don't birds compose music, which for all we know might well have dramatic and narrative elements? As for tower building, humans arguably are outmatched in ingenuity by termites. Not all types of termite beings infest houses. Some design magnificent structures thirty feet high, as well as mounds worth mimicking in our own construction. Termite towers, unlike skyscrapers, are integral to their surroundings and make life, including by creating oases in desert areas. Fungus-farming termites design mounds to function like breathing lungs, with carbon dioxide and oxygen exchange and regulated humidity. "We think humans are the best designers," engineer Kamaljik Singh observes, "but this is not really true."[63]

The February 2008 cover of *GSA Today* (Geological Society of America) features two towers dominating the megacity of Shanghai and poses the question: "Are We Now Living in the Anthropocene?" A 2011 article in *National Geographic*, "Enter the Anthropocene—Age of Man"[64] also shows a skyscraper, now dwarfing the megacity of Dubai. These photos bring to mind a *New Yorker* cartoon depicting two Euro-American male executives looking out an office window at the Empire State Building. Written down the skyscraper's face is the word VIAGRA. One of the admen quips: "Now that's product placement."[65] Everyone gets the joke because towers—along with guns, spires, scepters, swords, drills, and missiles—are common phallic symbols. Phallic symbols present power as dominance, the ability to penetrate the "mysteries of Nature," to "escape the gravity that is the earth's Force," and vanquish the "dark, chthonian female powers."[66] Skyscrapers signify men's separation from and rule over the "lower order" of the "mother, primal matter, the earth, and all that is conceived as belonging to it."[67] Architect Scott Johnson avers that there is "no artifact more synthetic and comprehensive . . . than the skyscraper," none more able to signify "expedience, transcendence, ambition, and dominance."[68] In other words, skyscrapers are phallic symbols, representing the idealized "apex of male power."[69]

Anthropocene discourse is full of verbal phallic signifiers like *potent, overwhelming, force,* and *dominance.* In traditional Catholicism, these all merge, again, into God the Father, whose signature title is *"The Omnipotent."*[70] Religious historian Jeffrey Kripal affirms what we might suspect: "Any prayer, devotional sigh, scriptural text, or sacred story that understands the divine as a 'father' is an implicitly phallic expression."[71] Less reverently, the heretical Mary Daly names this deity a "god/rod."[72]

Phallic symbols don't have anything, really, to do with what they supposedly represent—the biological penis. The penis is attached, embodied, sensitive, fertile, and fragrant. In reality the penis is far more flower than tower. The synthetic, detached, permanently hard (omnipotent), penetrating (but impenetrable), insensitive, straight-up, and straight phallic symbol is the emblem of *The Man.* In his world, *big* usually is better—but not always. Philosopher Susan Bordo explains that the phallus, which, after all, is an artifact, stands for *"generic* male superiority—not only over females but also over other species" and "the values of 'civilization' rather than 'nature,' with the Man who is made in God's image, not *Homo sapiens,* the human primate."[73] As such, Black men in particular are not identified in racist representations with that superiority-signifying phallus. Instead, they are projected as having large, hence "animalistic" penises, a stereotype also supporting the lie that they have a greater propensity to rape.[74]

Ancient phallocentric myths extol men's conquest of nature in stories of a sky god triumphing over a female earth deity. One of these, the influential "creation myth" told in the Sumerian *Enuma Elish,* provides "the mythological roots of all three patriarchal [Abrahamic] religions."[75] In this, the younger sky god Marduk slays the dragonish, land and sea-identified grandmother-creator Tiamat by directing an "evil wind" into her open mouth. He then tears her apart and forms a new world from her dismembered body. In some accounts, Marduk kills Tiamat with his penis.[76] This myth, one of the earliest accounts of a god or hero motherfucking Nature-Earth to death, was recited annually in ancient Sumer in order to "'assist' the victory of the god."[77] Today, this same basic story is told regularly, one way or another, in the continuing saga of Anthropocene Man. It retains that same function: to assist and renew the hero-god's victory and convince everyone that Mother Nature-Earth really *is* dead, reduced to the supposedly inanimate dirt upon which we tread.

The cover of *Scientific American* from December 2015 announces "How We Conquered the Planet"[78] and shows a facially indistinct, darkish-skinned man striding across the top of the planet holding phallic weapons. Although supposedly about prehistory and humanity's skills of cooperation, the picture tells a different story—the equation of the male with the human, of individualism, phallic weapons, and righteous dominion. This cover isn't really about prehistory. It is about the Anthropocenic now.

This image of a man towering over the planet is a visual cliché. Usually, it is White men and boys—conventionally masculine, straight, and wealthy-looking. They confidently hold Earth in their hands, spin the planet on their fingers, hold it with a string like a balloon, or pop the globe like a piece of bubble gum. They plant themselves like phallic flags, standing erect and victorious atop a diminutive planet.[79] These collectively recall W. Grancel Fitz's sardonic artwork *Big Baby* (1930), showing a huge White baby boy lording it over his tiny mother, who cowers in a high chair[80] (Figure 3.3). Fitz's clever work broadly illustrates psychoanalytic philosopher Teresa Brennan's argument that the masculine-identified, individualist "ego comes into being and maintains itself partly through the fantasy that it either contains or in other ways controls the mother; this fantasy involves the reversal of the original state of affairs, together with the imitation of the original."[81] That fantasy of omnipotence includes Big Baby fancying himself capable of killing, despoiling, overwhelming and even replacing the mother with no ill consequence to himself.

Related visuals in Anthropocene pop imagery and advertisements recurrently show young, slim, attractive women being violated in any number of ways—branded, strangled, dumped in the

FIGURE 3.3 W. Grancel Fitz, *Big Baby* (1930). Keith De Lellis Gallery, 16″ (height) × 12″ (width), photo i. gelatin silver.

trash, laid out as if dead, and fragmented into parts—with their bodies partially or wholly replaced with a product or a mechanism. Just as common are parallel images of the Earth being treated in much those same ways.[82] For example, a print ad for the Ford Fusion shows the car fusing into the face of a young Black woman;[83] the implication is that both are sexual objects to possess and drive. A planetary counterpart can be found on the cover of a crossword puzzle book, depicting Earth as composed partly of a puzzle, while an outsize pencil triumphantly penetrates the core.[84]

Paula Gunn Allen states what is obvious to Pueblo peoples: "We are the land, and the land is mother to us all . . . The earth is the source and the being of the people, and we are equally the being of the earth."[85] No one survives without Earth, not even *Man*. These normalized images of dominated, violated, dismembered, and replaced women, along with a planet treated in just those same ways, reveal not only "Big Baby's" matricidal and mass-murderous ideations, but ultimately his suicidal ones.

AS IT WAS IN THE BEGINNING . . .

Biblical references are common in Anthropocene discourse, usually invoking Genesis. The cover of Mark Lynas's *The God Species* incorporates familiar Sistine Chapel imagery—the extended fingers of the Father God and Adam. Between those reaching fingers is positioned a disproportionately small and round Earth, which is used to form the "o" in the title word *God*. The Genesis homage continues in the text as Lynas proudly proclaims: "On a planetary scale, humans now assert unchallenged dominion over all living things."[86] Of course, this is not nearly so true as Lynas would like. Astrophysicist Neil de Grasse Tyson offers a contrasting viewpoint bringing us back down to earth: "We are not in charge. Fecal bacteria are in charge!"[87]

Another Anthropocene text is *Regenesis: How Synthetic Biology Will Reinvent Nature and Ourselves*. Its title is not the only reference to the biblical creation story; the cover illustration features another piece of Genesis-inspired European art. Coauthors George Church and Ed Regis, a Harvard geneticist and a journalist, respectively, place synthetic biology at the fore of techniques to assure humans' ability to "remake nature according to their wishes."[88] Proselytizing the field's powers (particularly when fused with corporate capitalism), Church and Regis define *synthetic biology* as "the science of selectively altering the genes of organisms to make them do things that they wouldn't do in their original, natural, untouched state."[89] Their words are reminiscent of Columbus's plan for the Native peoples he first encountered: "With fifty men we could subjugate them all and make them do whatever we want."[90]

This fantasy of forced touching and coerced behavior bespeaks a sexually violent and slave-based model of interaction. Mary Daly denounces such "necrotechnology" as implementing a "rapist" model, "invading, poking, prodding, mutilating, destroying the core of living beings."[91] The *Regenesis* myth told by Church and Regis goes beyond *Man* forcibly touching, raping, conquering, and possessing all he defines into nature. It is a fantasy akin to Ira Levin's *The Stepford Wives*,[92] a feminist parable in which threatened husbands kill off their independent wives and replace them with perfectly obedient, home-making and sexually servicing robots. The parallel fantasy of some synthetic biologists involves knocking off Mother Nature and then installing an artificial substitute they control and force to produce and reproduce as they dictate.

The word *genesis*, like the word *nature*, derives from a Latin word meaning "birth" (*OED*). Yet the births in Genesis are bizarre and backward. The Father God makes the Mother Earth and Adam gives birth to Eve.[93] A popular feminist slogan riotously reverses that last absurdity: "I didn't come from your rib. You came from my vagina."

The story of Genesis as well as the future imagined by synthetic biology are part of a long history of philosophic, religious, and scientific traditions (typically but not always Western in origin) traced by philosopher Nancy Tuana, who calls attention to this recurring role reversal through which the male is identified as "the primary creative force."[94] Zora Neale Hurston, ever cunning, upends this same nonsense in *Their Eyes Were Watching God*. In the novel, a group of men, the porch sitters, debate the ultimate power in the Universe. One man, Lige, nominates "God," but another, Sam, argues for "nature," ending the debate with a powerful conclusion. Sam at first seems to be giving "God" the edge, but if you listen closely, he actually is affirming the primacy of nature: "It's nature . . . de strongest thing dat God ever made, now. Fact is it's de onliest thing God ever made. He made nature and nature made everything else . . . Nature is de first of everything."[95]

The absurd insistence on the primacy of the masculine, Mary Garrard points outs, persists even as the "reproductive power of both women and nature [is] everywhere apparent in the world."[96] The title of Jeremy Davies's *Birth of the Anthropocene* evokes female reproductive power, but the book itself effectively forgets about it. This, too, is a kind of subordination of nature and mothering power. Anybody or anything born comes from a mother— but in an implicitly phallic cosmology, the *birth* of eras, like the birth of nuclear bombs and AI "mind children,"[97] belong to men. Another envious reversal is applied to that other side of the mothering power, death, which becomes something that Anthropocene *Man* commits massively, while also aiming to control and ultimately defeat.

In *The Human Age*, Diane Ackerman profiles a number of key male thinkers and practitioners who represent what she sees as the positive aspects of the Anthropocene. One of these is scientist, inventor, and post-human philosopher Ray Kurzweil, who predicts "human" immortality and avows that "by the 2030s we'll be putting millions of nanobots inside our bodies to augment our immune system, to basically wipe out disease."[98] The aspirations to godhood are clear, but Kurzweil's "we" leaves a great deal unsaid. Even if this outcome were possible, do any of us imagine that this will be accessible to any but the most privileged, who also regard themselves as so worthy and exceptional that they should live forever?

WORLD WITHOUT END, AMEN

In his drawing "Earth Mother of Us All" (1901), Austrian artist Alfred Kubin registers a deep dread of the continuous rebirth that is nature. He depicts Earth as a monstrous, naked, pregnant yet older woman with long straggling black hair. She strides across a shadowy space with raised arms, sowing seeds and leaving a trail of skulls in her wake.[99] More than a century later, transhumanist philosopher and immortality enthusiast Max More, perhaps motivated by a similar dread, pens a rousingly defiant "Letter to Mother Nature." Transhumanists believe that humans can and should, with the aid of technologies, evolve beyond what they perceive to be the physical and mental limitations imposed by nature.[100] More, who represents himself as speaking for humanity, begins with condescension, expressing gratitude to "Mother Nature" for "raising us from simple, self-replicating chemicals to trillion-celled mammals" and for having "given us free rein of the planet."[101] Of course, neither claim is true. The first is a projection of human supremacy. The second is a justification for rape and conquest.

More then goes on to chide "Mother Nature" for doing a "poor job with the human constitution," including by making "humans vulnerable to disease and damage." While she has done "the best [she] could," More declares, her time is up. He hails a new adulthood for the human species, which includes refusing death: "We will no longer tolerate the tyranny of aging and death. Through genetic alterations, cellular manipulations, synthetic organs, and any necessary means, we will endow ourselves with enduring vitality and remove our expiration date. We will each decide for ourselves how long we shall live."[102]

More claims the transhumanist desire for immortality as a sign of maturity, but it seems more Big Baby egotism and insecurity. The "first great 'hero'"[103] of Western civilization is Gilgamesh, the ancient Sumerian demigod so obsessed with immortality that he committed ecocide, slaying a cedar forest's *daemon* Humbaba and then razing the forest to the ground. Scholar Robert Pogue Harrison notes that ancient texts describe Gilgamesh as a "builder of . . . walls"[104]—walls both material and psychical. Gilgamesh was supremely individualistic and obsessed with death, consumed with a "tragic and futile quest for personal immortality," full of "childlike rage," and possessing a "monumental will to power."[105] Nowadays the Gilgameshian figure seeking to build walls and live forever is all too familiar, while the climate change that is his hallmark will likely doom those cedar forests in a hundred years or so.[106]

Other immortality questers are the more extreme of the *biohackers*—do-it-yourself types who want to upgrade their own biology to super-human status through gene-editing techniques. In his *Transhumanist Manifesto*, Nikola Danaylov wants to jettison biology altogether: "Transhumanists have nothing to lose but their biology. We have immortality and the universe to gain."[107] Sounds like *Mas y Mas y Mas y Mas* all over again. Such efforts, from the perspective of eco-ability activists-theorists, seek a not only impossible but also undesirable "perfection."[108]

Worth: The Evolution of Financial Intelligence is the kind of top-drawer publication that includes ads for "His and Her" private jets. A 2015 issue reports on the male "Business and Tech Impresarios" now investing billions "in their next disruption: ENDING DEATH."[109] Little thought seems to be given to the capacity of the Earth to sustain even a tiny population of billionaire immortals, jetting privately around the world. Will the ecological footprint calculator[110] soon have to include "immortality" when estimating one's overall impact? Nor does there seem to be any consideration to the subsistence needs of other people, the pollution generated by this "disruption," and the routine torture of mice, dogs, and other beings in the medical research that accompanies these rich men's quests to increase their shelf lives. While billionaires deploy their wealth to ensure the health and safety of their own neighborhoods, foods, water, and air, nine million humans die painfully and prematurely every year from industrial contaminants, mostly air pollution.[111] I thought of all this while selecting a sympathy card for a friend, finding one showing verdant trees and quiet pools.[112] The message inside offers comfort in the knowledge that departed loved ones live on in the natural beauty that surrounds us.

Would those responsible for the Anthropocene make different choices if they faced not only their individual mortality, but also their collective fate, as the earthbound, to endure in the increasingly toxic conditions that surround us?

Another route to immortality favored by transhumanists entails downloading human consciousness into machines.[113] The insightful film *Get Out* (2017), written and directed by Jordan Peele,[114] points to the horror of this self-absorbed desire. A rich, young White woman seduces and entraps Black men and women in order to deliver them to her father, a neurosurgeon, and to her mother, a psychologist-hypnotist named "Missy" (as in plantation mistress), who head an organization of other rich, White people. Their mission is to abduct Black people and auction them off to their compatriots who will then be "downloaded" into these younger bodies, acquiring not only extended life but also the Black victim's particular gifts. The mother hypnotizes the victims, forcing their psyches into a horrific "sunken place" where they lose autonomy but retain awareness. The father performs the necessary brain surgery, putting the buyer's brain into the Black person's body, which the buyer now controls. The protagonist of *Get Out* is an art photographer bought by a blind and talent-less curator who wants to possess the artist's vision. *Get Out* exposes a master-slave model underlying the seemingly techno-utopian desire for enhancement and immortality. Anthropocene *Man* and his *Missy* would do well to heed the admonition given by blues singer Memphis Slim, "Don't care how great you are/ Don't care what you worth/ When it all ends up you got to/ Go back to Mother Earth."[115]

DEFIENDE NUESTRA MADRE TIERRA

When I was around twelve, I dreamt I was walking along a dirt path in a place I knew well, where branches stretched toward each other forming a canopy overhead. I heard troubling cries and looked to see three identical women bound to three contiguous trees. I felt their anguish as they called out to me in a language I did not understand. Still, I knew that they were telling me to act in their defense.

Around the world, many are hearing and heeding a terrestrial call to defend Mother Nature-Earth. During the 2014 People's Climate March in New York City, votive offerings and rituals devoted to the primal Earth deity were everywhere. A poster, created for the event by Favianna Rodriguez, depicted a brown-skinned, child-carrying activist, whom Rodriguez describes as a "fierce mama standing up for our mother, Pachamama, and defending her family and home."[116] The words on the poster were: *DEFIENDA NUESTRA MADRE* (DEFEND OUR MOTHER); Figure 3.4). One group of marchers carried a large puppet, a dark-skinned grandmother in traditional peasant dress, with accompanying signs reiterating the invocation: "*Defienda a Nuestra Madre Tierra.*" Another sign was emblazoned with a blue-green, black-haired woman with Earth at womb level and bearing this warning: "Don't mess with our Mama: Defend Gaia."

Global ecofeminist and environmental justice scholar-activist Vandana Shiva, a featured speaker at the march, has identified the stratagem behind colonizing cultures' deliberate devaluation of Mother Nature-Earth:

> The colonisation of diverse peoples was, at its root, a forced subjugation of ecological concepts of nature and of the Earth as the repository of all forms, latencies and powers of creation, the ground and cause of the world. The symbolism of Terra Mater . . . [inspires] ecology movements in the West today.[117]

A day prior to the People's Climate March, I interviewed Shiva and asked her to say more about why she regularly speaks of Mother Earth. Shiva replied, "Because she *is*."[118] Gonnella Frichner (Onondaga), president and founder of the American Indian Law Alliance, would agree, explaining that while "many

FIGURE 3.4 Favianna Rodriguez, *People's Climate Defend Our Home*, © 2018 Favianna Rodriguez. https://www.favianna.com.

non-indigenous people think the expression 'Mother Earth' is a metaphor . . . it is not."[119] Leroy Little Bear affirms: "Tribal territory is important because Earth is our Mother (and this is not a metaphor: it is real)."[120] Further supporting this awareness, Marilou Awiakta offered these canny comments: "Earth is not 'like' a mother; Earth *is* a mother. The Western way puts things at one remove. So, something no longer *is*, but only is *like*, becoming a metaphor."[121] This "one remove" mode is endemic to the phallic universe, where the artifact substitutes for the reality. *The Universal Declaration of the Rights of Mother Earth*, again, defines Mother Earth as "an indivisible, living community of interrelated and interdependent beings with a common destiny."[122] Not believing in this Mother is akin to not believing in climate change.

Legal efforts to establish the rights of Mother Earth have met with success in several places, including in Bolivia,[123] India, and New Zealand. In the latter case, a group of Maori people gained legal recognition for the rights of the river who is their family member; New Zealand must now officially recognize the river as a "living entity" with rights just like a human being. Gerrard Albert, the lead negotiator for the Whanganui iwi Maori, makes it clear that the river is alive—a being, an ancestor—and must be treated as an "indivisible whole, instead of the traditional model for the last 100 years of treating it from a perspective of ownership and management." The Whanganui iwi people's kinship with the river informs their recognition that "rather than us being masters of the natural world, we are part of it."[124]

Defense of a living river also motivated the 2016 gathering at Standing Rock, known as #NoDAPL. Scholar Kim TallBear (Sisseton-Wahpeton Oyate) identifies this movement as a women- and two-spirit centered effort, akin to #IdleNoMore and #BlackLivesMatter.[125] Numerous Dakota, Lakota, and Nakota, the Oceti Sakowin people, and an international contingent of allies gathered for almost a full year to protect Mni Sose, the Missouri River, from the Dakota Access Pipeline (DAPL). A statement from the Oceti Sakowin Camp of "Water Protectors" speaks to the need to defend land and water,

to ensure "the welfare of all people by honoring human rights, treaties, agreements, and cultures," and "to peacefully and prayerfully
defend our rights, and rise up as one to sustain Mother Earth and
her inhabitants."[126] Military veteran and Water Protector Brenda
White Bull (Lakota) spoke of the fight as the fulfillment of a prophecy delivered from elders (including her great-great grandfather
Sitting Bull), which foretold humankind gathering to defend Earth/
Nature. "It was called upon us to be here, to lead this, to lead this
fight, of protecting Unci Maka, our Grandmother Earth, and to protect our Mni Wiconi, our water of life."[127]

Jaskiran Dhillon and Nick Estes (Kul Wicasa, Lower Brule Sioux
Tribe) clarify the guiding vision of #NoDAPL: "Mni Sose is not a
thing that is quantifiable according to possessive logics. Mni Sose
is a relative: the Mni Oyate, the Water Nation. She is alive. Nothing
owns her. Thus, the popular Lakotayapi assertion 'Mni Wiconi'—
water is life or, more accurately, *water is alive*. You do not sell your
relative, Water Protectors vow. To be a good relative mandates protecting Mni Oyate from the DAPL's inevitable contamination. This
is the practice of *Wotakuye* (kinship) . . ."[128] It is imperative for everyone to know-feel that *Mni Sose* is not "like" a living being, but *is* a
living being. It has always been wrong to enslave a river; now, thanks
to the work of Indigenous activists, it is also, in some cases, illegal.

This grasp of spiritual-political reality was at the crux of the
work of murdered Honduran Indigenous rights activist Berta
Cáceres. Scholar Egla Martínez Salazar hails Cáceres as putting
"into practice grounded revolutionary spirituality," defending
Mother Earth, whom Cáceres called "Madre Naturaleza," and
fighting against "deeply interrelated patriarchy, colonialism and
racism." Cáceres never claimed, Salazar continues, to "giv[e] voice
to the voiceless," a common yet deeply colonial and racist refrain,
but instead introduced "the notion of lifting voices, lifting the
Lenca voice, knowledge and practice."[129]

For decades, Cáceres worked to prevent the construction of a
hydroelectric dam on the Gualcarque River. The Honduran company Desarrollos Energéticos led the dam project, accompanied

by strong international backing. In March 2016, a team of men in the service of this company assassinated Cáceres,[130] making her one of hundreds of land defenders murdered every year. Honduras is the most dangerous place in the world for environmental activists[131] after more than a century of international interference and extraction of the country's "natural common goods"—Cáceres' phrase repudiating the possessive and objectifying language of "'natural resources.'"[132] Grahame Russell, from the non-profit group Rights Action, concludes that Berta Cáceres "was killed by all those people, countries and institutions whose greed and violence she lived, stood and fought against. Berta lived against all injustices, all inequalities, all discriminations, all Mother Earth destroying activities."[133]

In 1993, Cáceres co-founded COPINH, The Civic Council of Popular and Indigenous Organizations of Honduras, an organization made up of two hundred Lenca communities in the western part of the country. Its continuing purpose is to defend Lenca rights, territory, "and our natural resources as part of our Lenca cosmovision of respect for Mother Earth."[134] COPINH is explicitly anticolonialist and anti-heteropatriarchy, supporting women's and LGBTQ rights and recognizing these as inseparable from environmental-justice concerns.

In 2015, Cáceres, accepting the Goldman Environmental Prize for her campaign against the hydroelectric dam, made it plain that she acted in response to the Gualcarque River's "call" and to Mother Earth's "demand":

> In our worldviews, we are beings who come from the Earth, from the water and from corn. The Lenca people are ancestral guardians of the rivers, in turn protected by the spirits of young girls, who teach us that giving our lives in various ways for the protection of the rivers is giving our lives for the well-being of humanity and of this planet. . . . We must shake our conscience free of the rapacious capitalism, racism and patriarchy that will only assure our own self-destruction. The Gualcarque River has

called upon us, as have other gravely threatened rivers. We must answer their call. Our Mother Earth, militarized, fenced-in, poisoned, a place where basic rights are systematically violated, demands that we take action. Let us build societies that are able to coexist in a dignified way, in a way that protects life. Let us come together and remain hopeful as we defend and care for the blood of this Earth and of its spirits.[135]

The guiding vision that has produced the Anthropocene ordains activities like the blood-poisoning of land and rivers as *rational.* At the same time, it scorns as folkloric and fantastical the recognition of Mother Nature-Earth as real, ridiculing and marginalizing people who acknowledge hearing and responding to the calls of trees, rocks, other animals, water, and soil. Those listening to Cáceres's words but mired in the Western way of "one remove," even if they agree with her politics, might categorize her talk of "spirits," "calls," and "Mother Earth" as either nonsense or metaphor. This would be wrong.

When I refer to Mother Nature-Earth in academic settings, even as I stress nonviolence, anti-racism, and feminism, and even as I draw upon Indigenous authors and activists to explain this reality's contemporaneity and complexity—including gender complexity—I am often dismissed by other White feminists as essentialist, fluffy, simplistic, and even as employing dangerous blood and soil fascist-type rhetoric. There is indeed a strain of nature writing and environmentalism that is anti-immigrant, anti-Semitic, ableist, misogynistic, homophobic, nativist, and White supremacist.[136] But this, precisely, is what environmental-justice-based invocation of Mother Nature-Earth (whose essence is commonality amidst difference) challenges. An eco-ability perspective rightly insists that "difference . . . is the essential ingredient for human and global survival."[137]

A rebuke against speaking of the feminine and the land, Vanessa Watts writes, is sometimes directed at Indigenous scholars. Watts traces this to the colonizing culture's "corruption" of "essential categories of Indigenous conceptions of the living, thinking and speaking world (the feminine and the land)."[138] This is not simply a matter of philosophical debate, but a deliberate stratagem enabling conquest and possession:

> What happens when soil is removed from territory? What happens when flesh is taken from the body? More importantly, what happens to the territory after its resources are excavated? Shopping malls and paper mills —a literal excavation of thoughts are forcibly transformed into objects of the colonial imperative. Those crops became their crops, that tree became their trees and so on and so on. Once the voices and thoughts of these two essential categories of creation (the feminine and land) are silenced and then corrupted, the acquisition and destruction of land becomes all the more realized.[139]

Paula Gunn Allen, in *The Sacred Hoop: Recovering the Feminine in American Indian Traditions*, writes of the thinking "feminine" or "female" as "context" itself and as "God, because it is the source and generator of meaning."[140] As Allen further explains, there is no binary and hierarchical gender system underlying this; rather, the feminine includes the masculine, with the masculine understood as "a subset of the feminine."[141] Putting Allen and Watts in conversation with Hurston, Mother Nature-Earth is the "the first of everything" because the feminine is needed for ongoing *conception*—a word linking thought and fecundation, creativity and procreativity.

This *feminine* for Cáceres, Watts, and Allen also has nothing to do with the Euro-Western heterosexist ideology of women as a fixed, biologically inferior sex class. Watts observes that "Euro-Western discourses" critiquing essentialism are rightly attempting to "remedy historical mistakes of biological essentialisms (i.e. scientific racism)." Still, she continues, it is downright wrong to measure "Indigenous cosmologies . . . against the products of Euro-Western mistakes. Nor should Indigenous peoples be the inheritors of these mistakes." If Indigenous people "disengage" from "essentialism," Watts avers, they "run the risk of disengaging with the land."[142]

Watts points to the biblical origin story, Genesis, as disastrous in its disconnection of people from land, its dis-spiriting of other-than-human nature, and its condemnation of the feminine in the persona of Eve, the Mother of All the Living. Eve "becomes responsible for all the pain of childbirth and resentment for being cast out of paradise. The interaction of Eve and the Serpent results in shame and excommunication from nature. Additionally, future dialogue and communication with animals become taboo and a source of witchcraft. It is at this point of conflict where thought, perception, and action are separated from the supposed inertia of nature."[143]

Genesis is an origin story not of the world but of heteropatriarchy, which nonetheless claims itself to be the world.[144] Still, by some twist of fate, Genesis can't help but communicate the means for unmaking that world.[145] Eve's supposed "sin" was listening to the Serpent-Nature, conversing, accepting, and sharing the Mother's gift—just as Rachel Carson, Robin Wall Kimmerer, Berta Cáceres, and all "extra keen" perceivers recommend. The real sin is to be insensitive to the presence and awareness of others, to take and steal their gifts, believing it is your due as the superior and singularly sentient species, to refuse to listen, and to neglect humans' responsibility to land.

WORSHIPPING THE GROUND WE WALK ON

On the first year anniversary of Berta Cáceres's assassination, COPINH issued a collectively composed poster (Figure 3.5).[146] It shows her smiling face in profile, with her long black hair cascading down in the shape of the land of Honduras. Colorful trees and flowers grow from her, the river flows from her; a few buildings arise from her, and COPINH marchers process along her, holding signs of their commitment. The poster's text reads: "A 1 AÑO DE SU

FIGURE 3.5 Berta Cáceres COPINH poster (2017). Consejo Cívico de Organizaciones Populares e Indígenas de Honduras.

SIEMBRA BERTA VIVE COPINH SIGUE" ("On the first anniversary of the sowing, Berta lives, COPINH follows").[147]

One interpreter relates a defiant Latin American proverb—"They tried to bury us, they didn't know that we are seeds"[148]—to draw out the significance of those words. "My mother," Cáceres's daughter Bertita Zúñiga affirms, "is a seed that has multiplied."[149] Even though she was murdered, Cáceres's spirit takes root and grows and others continue her work. This poster shows Cáceres as the ground of the movement as well as the ground itself; love for Cáceres fuses with love for land.

The Oglala Lakota Chief, Luther Standing Bear, spoke of the profound experience for human beings of direct contact with Earth: "The old people came literally to love the soil and they sat or reclined on the ground with a feeling of being close to a mothering power. It was good for the skin to touch the earth and the old people liked to remove their moccasins and walk with bare feet on the sacred earth."[150] Political leader and activist Faith Spotted Eagle (Dakota) concurs, invoking as well the specific sounds of Earth: "You know, Mother Earth has a sound. And she has a mothering sound. And if you're real quiet you can hear that. Everything on the land is a teacher."[151]

The Anthropocene's paradigmatic visualization is the sky-scraping, steely, concrete, plasticized and paved-over megacity, with thoroughfares traveled by racing people and machines. This megacity (mirrored in the prison industrial complex and indus-trial animal farming)[152] seems designed precisely to block tan-gible, heartfelt connections—tactile and audial—to land, to the "mothering power."

Compare this Anthropocene visualization to an artwork by Erin Marie Konsmo (Métis/Cree) that invokes the necessity of being in touch with Earth—and this is not a metaphor. A young Native woman crouches down and places one hand on the ground. Her body is within a circle. Above are the words: "Land Is Ceremony."[153] Ceremonies fuse the diverse beings of land into a whole. Ceremonies help everyone to "'remember to remember'"[154] the interconnection, awareness, thought, agency, and oneness of all. Mother Earth in this way is an "altar."[155] Happening on the creative and procreative altar of Earth all the time is the cere-mony that *is* land—the cyclical and renewing process of ongoing creation—life, death, and rebirth. Human participation in that ceremony includes knowing that "when you put your hand on that earth, you should feel something pushing back."[156] Human par-ticipation in that ceremony constitutes veritably worshipping the ground we walk, sit, roll,[157] lie, or stand on. This includes honoring treaties and returning stolen land.

Scholars Eve Tuck (Unangax̂) and K. Wayne Yang abhor deliberate interference with Indigenous people's participation in the ceremony of land:

Within settler colonialism, the most important concern is land/ water/air/subterranean earth (land) . . . Land is what is most

valuable, contested, required. This is both because the settlers
make Indigenous land their new home and source of capital, and
also because the disruption of Indigenous relationships to land
represents a profound epistemic, ontological, cosmological violence. This violence is not temporarily contained in the arrival of
the settler but is reasserted each day of occupation."[158]

"Decolonization," Tuck and Yang declare, "is not a metaphor."
Decolonization, they say, is not abstract, nor is it equivalent to
consciousness raising. It requires actions, including honoring
treaties and returning stolen lands to Indigenous peoples so that
they can renew disrupted relationships with land.

Decolonization requires ending racism and paying repara-
tions to the descendants of enslaved peoples, putting a stop to
rape in all forms, and, in all ways, refusing to feed those gods of
"speed, quantity, and the quick and easy profit." The "monstrous
evils" wrought by worship of those gods, Carson warned, "go
long unrecognized" because "those who create them manage by
some devious rationalizing to blind themselves to the harm they
have done society."[159] Many of us participate in those harms by
our privileges and the lifestyle that comes with them, often falsely
believing, like those White buyers of Black people in *Get Out*, that
this is our due.

Putting himself at that "one remove," materially and figu-
ratively, *Man*, with cooperation from *Missy*, attempts deicide of
Mother Nature-Earth, all the while worshipping himself. At the
same time, those on the ground hear the "mothering sound"
(Spotted Eagle) and the cries of Earth, feel the trouble, know the
pain of the traumatized and violated ground as their own pain,
and answer the call to interrupt that assault. *Man's* rape of the
Earth is not a metaphor.

THE ANTHROPOCENE IS

A MOTHERFUCKER

AN ONGOING DEBATE CONCERNS THE Anthropocene's origins.[1] Did it begin fifty thousand years ago when humans first
started burning forests deliberately? Or was its inception the
invention of plow agriculture or, later still, nuclear weaponry? If
you believe the Anthropocene began at the moment when humankind first exhibited any kind of environment-changing behavior, it might well be fifty thousand years old. But if you consider,
instead, that the Anthropocene really is the *Age of Man*, then the
era began with the advent of heteropatriarchal systems and the
institutionalization of motherfucking.

Gerda Lerner, the founding mother of the field of women's
history, argued that the "creation of patriarchy" occurred when
some men first seized control over and commodified female sexuality and powers of maternity and then went on to invent a social
hierarchy based on difference that ultimately established slavery,
private property, and class society.[2]

Ecofeminist thinker Susan Griffin, writing in 1971, was one
of the first to point to a correspondence between men's sexual violence against women and environmental devastation. She begins
by advancing an understanding of a *rape culture*, one that institutes rape as a means of social control, asserting and affirming
male superiority, while also becoming a paradigm for abusive
practices: "Rape is not an isolated act that can be rooted out from

Call Your "Mutha'". Jane Caputi, Oxford University Press (2020). © Oxford University Press.
DOI: 10.1093/oso/9780190902704.001.0001

patriarchy without ending patriarchy itself. The same men and power structure who victimize women are engaged in the act of raping Vietnam, raping Black people and the very earth we live upon . . . As the symbolic expression of the white male hierarchy, rape is the quintessential act of our civilization."[3] Scholar Jack Forbes concludes much the same in his condemnation of settler colonialism in the United States, *Columbus and Other Cannibals*: "The rape of a woman, the rape of a land, and the rape of a people, they are all the same. And they are the same as the rape of the earth, the rape of the rivers, the rape of the forest, the rape of the air, the rape of the animals."[4]

Martin Phillips, reflecting on the horror of being raped by a male acquaintance when he was a college student, also implicates more than his individual rapist: "A society that punishes with violence, finds justice in violence, enjoys violence as entertainment, mixes sex with violence, feeds itself by violently killing animals, and rapes the earth as its daily business is a mean motherfucker."[5] The Anthropocene is an expression of just such a society.

To further comprehend the origins and meaning of the Anthropocene, I begin with Lerner's research into the introduction of the specific form of heteropatriarchy that still holds sway in the Euro-Western world. Next, I consider the United States, the exemplar if not the architect nation of the Anthropocene, and the veritable motherfucking attending the genocide and chattel slavery at its founding. These atrocities made possible the current wealth and status of the United States and hence contributed immeasurably to the making of the Age of *Man*.

THE ORIGINS OF *MAN*

Probably many of us received an early pop-culture lesson in heteropatriarchy from a cartoon—the ubiquitous one of a White caveman beating a blonde, curvy cavewoman over the head with a club and then dragging her back to his cave. The implication is that he will rape her, keep her, and that she will serve and service him, including by bearing and nurturing "his" children. This cartoon fuses sex and violence, while also making male supremacy, rape, battery, sexual slavery, and forced reproduction seem natural and even inventive (after all, the man uses a tool). With a shrug and a laugh, it says that this is the way it always has been and always will be. In my nearly forty years of university teaching, I have regularly encountered students who confidently assert that male domination is inevitable because "it goes right back to the cavemen." Gerda Lerner, founding mother of the discipline of women's history, says a firm "No" to this nonsense. Patriarchy does not characterize all cultures.⁶ Patriarchy has a beginning, and it will have an end (albeit perhaps an apocalyptic one).

In her 1986 work, *The Creation of Patriarchy*, Lerner examined the origins of the particular version of the male dominant system that developed in the Near East some seven thousand years ago and become the basis for Western civilization. Lerner began, she says, with the "usual economic questions" but soon realized that these did not provide satisfactory insight. She shifted her focus to sexuality, specifically the way that men gained "control of women's sexuality and procreativity" via such means as men's exchange of women,⁷ the sale of brides, slavery, prostitution, codes of female virginity, forced reproduction and the institutionalized heterosexist marriage and family, and the domination of younger men by older men. Women "themselves became a resource, acquired by men much as the land was acquired by men."⁸ After subjugating their own women, the patriarchal founding fathers went on to enslave first the women and children and then the men of "othered" outside groups. This combined, Lerner writes, "both racism and sexism."⁹

Women were the first form of property, their social status forcibly changed from autonomous being and source of life to controlled resource. The ramifications of this shift are felt to this day. This is why patriarchal customs prescribe that children take their father's name and married women (including child brides) take their husband's. It is why marital rape, still legal in some US states up until 1996 and in many countries to this day, remains difficult to prosecute,[10] why access to birth control and abortion continues to be stigmatized and restricted, and why so many men believe it to be their inalienable right to have sexual access to and control (whether personal, institutional, or state) over women's, girls', and boys' bodies.

Patriarchies, even though biologically inaccurate and in contradiction to observed realities,[11] divide everyone into either a *man* (said to be innately powerful, dominant, and aggressive) or a *woman* (said to be innately weak, corrupt, irrational, submissive, and passive). Mandated and unequal heterosexuality as well as rape, Andrea Dworkin argues, are inherent in those roles and definitions.[12] Patriarchs demand chastity and fidelity from "their" women, while reserving a class of public women, girls, and also (though this may be more hidden) boys for their general sexual use. Heteropatriarchy then and now relies on both social conditioning and force. The latter includes men's gender-based violence, including sexual violence, against women, girls, younger men and boys, LGBTQ people, and those who are unable and/or unwilling to conform to the standardized gender roles and expressions. System maintenance mandates a father-headed hierarchical family structure, a complicit legal code, economic dependence, and, finally, the enthronement of a heavenly father god-creator and concomitant overthrow of prior Earth-based religions, These earlier religions were based in "metaphysical female power, especially the power to give life," often represented with images of a naked goddess and amulets in the shape of the vulva.[13]

Lerner emphasizes that patriarchy developed gradually over millennia and that women, more or less, participated. The system divided women against each other in a top-down schema of class, racial, ethnic, religious, and other hierarchies. Within these hierarchies, women gained or lost social status based on their affiliations with men. Some were accorded the status "'respectable' (that is, attached to one man)." Below them were women deemed "'not-respectable' (that is, not attached to one man or free of all men)."[14]

For high-status women, the patriarchal set-up made them subject to legal rape, assault, abandonment, and sometimes even murder by the men in their families. Still, when things went relatively well, they had wealth, authority over enslaved peoples, and at least some protection from abuse by men not of their families. The upper-tier women were promised, in exchange for their cooperation, a secure future for themselves and their children. During the original colonization of the United States and Canada, "respectable" women allied with their men to gain status and power, including by helping them to impose male-supremacist norms of sexuality, gender, and family.[15] For example, Writer-activist Leanne Betasamosake Simpson (Michi Saagig Nishnaabeg) points to the imposition of heteropatriarchal values and practices on Michi Saagig Nishnaabeg peoples in what is now Canada. Methodist missionary White women took on the job of "policing" Indigenous peoples' bodies, sexualities, "gender variance and fluidity," births, family structure, and the education of children. These "White women," Simpson writes, "were out to destroy our intelligence and political systems. This is genocide. This is sexual and gendered violence as a tool of genocide and as a tool of dispossession."[16]

This patriarchal pattern can also be seen in Southern US plantation life, vividly depicted in the Oscar-winning 2013 film *12 Years a Slave* based on Solomon's Northrup's 1853 memoir. A White master exercises total command over all on his estate, including by raping enslaved women and girls and by emotionally abusing his wife. Meanwhile, his wife collaborates in his domination over everyone else, especially in his brutalization of Patsey, the woman he has singled out for his most extreme tortures.

Margaret Atwood's novel *The Handmaid's Tale*[17] is structured around the same basic pattern, set in a near future in which environmental contamination has resulted in widespread human infertility. A right-wing religious takeover of the United States puts an elite group of male "commanders" on top; these commanders legislate the sexual and reproductive enslavement of fertile young women. Each commander gets a "handmaid," whom they rape, supported by their privileged if subordinated wives—who also collaborate in the legal theft of the handmaid's children. A class of missionary-type older women known as the "Aunts" are in charge of teaching the handmaids, which means breaking their spirits, disciplining them, and torturing them. Of course, all this, as essayist and poet M. NourbeSe Philip makes plain, is not really dystopic projection, but the history of chattel slavery: "*The Handmaid's Tale* but a vision of the future past in white face."[18]

As I write, heteropatriarchy, White nationalism, anti-immigrant ethnocentrism, homophobia, transphobia, attacks upon women's sexual and reproductive rights, and environmental injustices are surging following the election of Gilgameshian wall-builder and accused rapist[19] US President Donald Trump.[20] Critics speak of a disastrous "Trumpocene."[21] On January 23, 2019, newly elected Michigan Congresswoman Rashida Tlaib just outright called President Donald Trump a "motherfucker,"[22] a historically resonant insult.

During March 2016, in the midst of the primary campaign leading up to Trump's victory, I had a conversation across a class and gender divide with a White man behind a fish-store counter in Long Island (New York). I told him that I didn't like Trump (at the time just one of several candidates) because of his hostility toward women, Muslims, Mexicans, and others. The man listened to this without argument, but went on to say that nonetheless Trump had his vote because Trump was just like the "men who founded this country." I didn't disagree.

SEX-AND-VIOLENCE
AMERICAN STYLE

Sex-and-violence is the signature of American popular culture.[23] The American-style fusion of the two derives from the nation's founding in the genocide of Indigenous peoples and chattel enslavement of African peoples. Having subordinated Mother Nature-Earth in Europe—ideologically as well as via deforestation, extirpation of wild animals, witch burnings, and the closing of the commons—conquest-minded European men sought new frontiers to continue the plunder.

These Europeans represented themselves as civilized, outside, and superior to nature, measuring manhood in possessions that included women, the enslaved (Africans and Native peoples), animals, and land. A characteristically American notion equated freedom with untrammeled individualism, unlimited mobility, righteous frontier-pushing, risk-taking, and profit-making. This ideal endures in the common understanding of violence as regenerative, sexy, and virile.[24]

American studies scholar Andrea Smith examines the "project of colonial sexual violence," which "establishes the ideology that Native bodies are inherently violable—and by extension, that Native lands are also inherently violable."[25] Legal theorist and anti-rape activist Sarah Deer (Muscogee) also perceives an intrinsic connection between the sexual violence attending and assisting conquest and the concomitant environmental violence. In *The Beginning and End of Rape*, Deer demands understanding that the invasion continues today as Native women are more than two times more likely to suffer sexual assault than their White counterparts and are most often abused by White men.[26] Exceptionally high rates of rape, battery, and sex trafficking occur in areas where extractive industries operate and "man camps" of outside workers are set up. Deer perceives that "the crime that Native women are experiencing as a result of the exploding fracking business has parallels with the harm being done to the planet—the land and water are being poisoned as the hearts and spirits of Native women break."[27]

This ongoing rapism is strategic in multiple ways. Paula Gunn Allen perceives that the conquest of Indigenous peoples included "a gynocidal motive behind the genocide"—the deliberate destruction of the often woman-centered, non-heteropatriarchal, noncapitalist, nonhierarchical, and ecologically healthy Native cultures.[28] These cultures in what is now North America, though diverse and by no means uniform, were not utopic ideals, but they also were *not* characteristically male dominant, compulsorily heteronormative, child-abusive, or culturally rapist. Sarah Deer and Liz Murphy (Mille Lacs Band of Ojibwe), write that most of these Native belief systems did not "historically put men in complete dominion over their property—namely women, children, and animals."[29] Accordingly, their cultures did not normalize men's sexual violence against women and children nor the maltreatment of other animals, plants, elements, and land.

Conquerors sought to erase Indigenous societies in order to impose heteropatriarchy and capitalism. Leanne Betasamosake Simpson explains that "Indigenous bodies, particularly the bodies of 2SQ people, children, and women, represented the lived alternative to hetero-normative constructions of gender, political systems, and rules of descent. They are political orders. They represent alternative Indigenous political systems that refuse to replicate capitalism, heteropatriarchy, and whiteness . . . and so it is these bodies that must be eradicated—disappeared and erased into Canadian society, outright murdered, or damaged to the point where we can no longer reproduce Indigeneity."[30]

There is a long history of Native resistance,[31] including specifically to rapism, Deer introduces readers to Sarah Winnemucca, a Paiute activist who exposed Euro-American men's sexual violence against Native women in her autobiography, *Life among the Paiutes* (1883).[32] Throughout the era of settler colonialism, Indigenous peoples continue to act for "rematriation" of land.

Steven Newcomb (Shawnee/Lenape), executive director of the Indigenous Law Institute, explains: "By 'rematriation' I mean 'to restore a living culture to its rightful place on Mother Earth,' or 'to restore a people to a spiritual way of life, in sacred relationship with their ancestral lands, without external interference.' As a concept, *rematriation* acknowledges that our ancestors lived in spiritual relationship with our lands for thousands of years, and that we have a sacred duty to maintain that relationship for the benefit of our future generations."[33] Settlers, including myself, can take it up as their duty to support the rematriation process in appropriate but not appropriative ways.

Indigenous societies did not "vanish," as the settler narrative would have it, but continue today, also providing alternative models. Congresswoman Deb Haaland (Laguna Pueblo), elected in 2018 from New Mexico, reminds us of this in her rebuke of Donald Trump's racist anti-immigrant policies, deeming these a sure indication that he understands neither his own place, nor the country: "As a 35th-generation New Mexican and a descendant of the original inhabitants of this continent, I say that the promise of our country is for everyone to find success, pursue happiness and live lives of equality. This is the Pueblo way. It's the American way."[34]

Along with genocide against Indigenous peoples, the consanguineous "New World" founding atrocity was chattel enslavement of Africans. It, too, combined parallel violences against bodies and land, while suppressing and discrediting alternative ways of thinking and being. In *Rooted in the Earth: Reclaiming the African American Environmental Heritage*, Dianne D. Glave affirms the differences between African ecological understandings and European ones: "Broadly speaking, Africans believed in the interconnectedness of the human, spiritual, and environmental realms and felt that harm toward or care for one necessarily affected the others."[35] Ecowomanist theorist Delores Williams applies that knowledge to her analysis of the planation master's ownership, rape, and forced reproduction of enslaved women: "Put simply," she avers, "the assault upon the natural environment today is but an extension of that assault upon black women's bodies in the nineteenth century." The plantation system's violation and exhaustion of land directly paralleled the masters' "defilement" of Black women in rape, forced reproduction, and "spirit-breaking."[36] That same assault, Williams perceives, continues today in environmentally destructive practices such as strip mining, which is currently used in Canada for tar sands oil extraction with grievous consequences for all the area's local beings, including its Indigenous communities.[37]

The plantation model, as analyzed by Williams, as well as by Angela Davis, Barbara Omolade, Dorothy Roberts, Adrienne Davis, Shateema Threadcraft, and Beth Richie,[38] among others, exemplified the extremes of violation of integrity, including racialized sexual abuse of women and girls, men and boys[39] and with ramifications continuing through the present day. Political scientist Threadcraft writes of the centuries-long White supremacist denial of "intimate justice" to Black women, including through definition of "the black female body as a body that invited such violations of integrity–as a body with no such integrity."[40]

Legal scholar Adrienne Davis speaks to the plantation system as a "sexual economy," one that "extracted from black women another form of 'work' that remains almost inarticulable in its horror: reproducing the slave workforce through giving birth and serving as forced sexual labor to countless men of all races."[41] The system, she says, "expropriated black women's sexuality and reproductive capacity for white pleasure and profit."[42] Masters and missies furthered the horror by making Black women care for White children, at the expense of their own. Owners enslaved, abused, and sold off those children, including the masters' own progeny born to Black mothers. The White master was the consummate bad father, although racism continues to lay this stereotype on men of color. Again, it was the master's biracial enslaved children, Black folk theory tells, who invented the word *motherfucker* to speak to the full dimensions of his crime.

Despite this widespread suffering, scholar Frances Foster critiques the destructive stereotype of enslaved Black women as "ultimate victims."[43] This stereotype is still held, Janell Hobson cautions, by some White feminists.[44] Black feminists model documenting and analyzing the horror, naming the perpetrators while also foregrounding women's resistance[45]—using contraceptives, fighting off and even killing rapists (including their masters), negotiating for what they could, and, like Margaret Garner, committing infanticide rather than see her daughter returned to enslavement. Many also were able to retain, in historian Daina Ramey Berry's expression, their "soul value"—their vitality of spirit, including self-love—despite the spirit-breaking intent of enslavers, slavery supporters, and slavery enforcers.[46]

Writing about the rapist plantation system and the systemic rape of Indigenous women by colonizers, I aim to avoid that "ultimate victim" stereotype, while also reckoning with the continuing significance of these horrors to the current realities of the Anthropocene. Thoughtful critics propose more politically accurate names for the era, including *Capitalocene* and *Plantationocene*.[47] These are helpful, but they don't account for the racialized, classed, gendered, and sexual domination that underlies the Age of *Man*. It was specifically the rapist genocide of Indigenous peoples and subsequent theft of their land, combined with specifically rapist chattel slavery, that enabled a worldwide cotton industry, which in turn made possible what Cáceres names global "rapacious capitalism"[48]—the system most responsible for the Anthropocene.

Scholar-activist Lee Maracle (Soh:lo Nation) muses: "The combined knowledge of African ex-slaves and colonized Natives in North America is going to tear asunder the holy citadel of patriarchy. Who can understand the pain of this land better than a Native woman? Who can understand the oppression that capitalism metes out to working people better than a Black woman? The road to freedom is paved with the intimate knowledge of the oppressed."[49]

Too often, Anthropocene analysts do not recognize this accumulated and combined knowledge. So too, in scores of books on the era there is little mention of gender-based violence factoring into the foundation of the Anthropocene, even in feminist analyses. But motherfucking cannot remain unmentionable. For what is the Capitalocene but *Man's* rape of all he deems *nature* as his "daily business" (Martin Phillips)? What is the Plantationocene if not *The Man's* motherfucking of all he deems *nature* for both "pleasure and profit" (Adrienne Davis)?

OVER HER DEAD BODY

Sometime after moving to Florida, in 1997, I had a disturbing dream. A kind mentor from graduate school appeared, uncharacteristically wearing a military uniform. He took me somewhere north and west in Florida and somberly pointed to the ground. Just underneath the surface were the bodies of countless murdered women. All land is composed of the corpses of ancestors, human and other than human. Death, like birth, is essential to the continuous cycle that is nature. But this dream was not about that. These women had been tortured, silenced, and dumped.

I'm sure this dream says something about the "cut-up land"[50] that is now Florida, a place civil-rights and labor activist Delores Huerta calls "ground zero for what's happening right now in the United States."[51] The state's history is marked with misogynist, homophobic, transphobic, and racist murders. It is a hub for coerced sex work, immigrant farmworker exploitation,[52] communities gated against those perceived as "other" (like Trayvon Martin),[53] shopping malls, and notorious mass shootings, as well as the home of the endangered and unique Everglades, a panther species on the cusp of extinction, famous theme parks, and the military headquarters for the United States Central Command. Thinking on the military garb in the dream, I recalled lines from an Air Force camp song: "I wish that all the ladies/ were holes in the road/ and if I was a dump truck/ I'd fill them with my load."[54]

In patriarchal cultures, writer Alison Fell remarks, "All women have imprinted in them the basic politics of male territoriality."[55] "Ladies"—with variations depending on intersecting factors of race, ethnicity, class, and sexuality—are treated as "holes" out of which energy and raw materials can be extracted and into which waste can be dumped—in a word, a "cum-dumpster."[56] Conquerors "territorialize" land. The same is done to enslaved people, scholar Katherine McKittrick explains, with bodies "territorialized . . . claimed, owned, and controlled by an outsider."[57] *Territory,* from *terra,* the Latin word for Earth, is Earth killed into property, cut up, claimed, developed, excavated, mined, mapped, occupied, used, and used up.

A self-promotion by *Hustler* magazine, from 1977, the calculatedly outrageous porn magazine, lays this out (Figure 4.1) Using conceptual and visual themes associated with the European scientific revolution,[58] the display places an unconscious, young White woman flat on her back on a hospital operating table. Three White men, costumed as gynecological surgeons, peer between her legs and compare what they see to a *Hustler* centerfold. The copy reads in part: "*Hustler* exposes unexplored territories and shows you parts of a woman's body you thought were only visible during a hysterectomy operation." Along with the standard lesson in rapism, viewers get a related one in epistemology.

Legal theorist Catharine Mackinnon proposes a "feminist theory of knowledge" that is "inextricable from the feminist critique of power."[59] Ruling men have set themselves up as the "knowers, mind," with women correspondingly put in the role of "'to-be-known,' matter, that which is to be controlled and subdued, the acted upon."[60] Accordingly, MacKinnon summarizes, "Sexual metaphors for knowledge are no coincidence. . . . Feminists are beginning to understand that to know has meant to fuck."[61] The "doctors" in the *Hustler* promotion represent that masculine "mind" in relation to supposedly mindless feminine

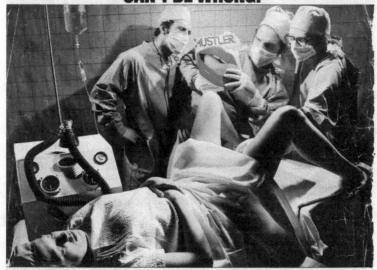

3.5 MILLION "AMATEUR GYNECOLOGISTS" CAN'T BE WRONG.

Now you, too, can enter the exciting world of gynecology. And you don't have to put up with eight years of medical school and skyrocketing tuition rates.

For just $22 you can subscribe to HUSTLER and learn everything about female anatomy. HUSTLER exposes unexplored territories and shows you parts of a woman's body you thought were visible only during a hysterectomy operation.

When you subscribe to HUSTLER you'll save $5.75 off the newsstand price and receive your copies two weeks in advance. Plus, each copy of HUSTLER is sealed in a black plastic wrapper, so no one can see it before you do.

Have HUSTLER delivered to your door. It's the only magazine that entertains you and teaches a course in gynecology at the same time.

Subscribe to HUSTLER

HUSTLER • P.O. Box 2204 • Columbus, Ohio 43216

☐ 1 year at $22 (U.S.) ☐ 1 year at $28 (FOREIGN)
☐ 2 years at $42 (U.S.) ☐ 2 years at $54 (FOREIGN)
☐ 3 years at $61 (U.S.) ☐ 3 years at $79 (FOREIGN)
☐ New Subscriber ☐ Renewal
50% discount to all U.S. servicemen overseas.

Please Print   HU1177

Name

Address

City, State, Zip
Enclosed is my ☐ check ☐ money order (cash not accepted), or charge to my ☐ VISA ☐ MC:

Numbers No.   Exp. Date
                                    mo   year

Signature, Date

EXPRESS CHARGE CARD ORDERING . . . 24-hour toll-free service. Order now by calling 1-800-848-9107. (In Ohio, call: 1-800-282-9216.)

FIGURE 4.1 "3.5 Million 'Amateur Gynecologists Can't Be Wrong,"

© Hustler Magazine Inc., November, 1977.

nature—sexually subjugated, silenced, and open to forcings, penetration, extraction, and mutilation. That last assault is suggested by *Hustler*'s reference, not only to "unexplored territories," but also to the "hysterectomy operation" about to be performed.

This display of the unconscious woman simultaneously updates the porno-tropical (McClintock) trope of *terra nullius*, whereby invaders, despite manifest evidence to the contrary, define the lands (*terra*) they desire as "empty" (*nullius*), belonging to no one, vacant physically as well as mentally and spiritually.[62] Accompanying this legalistic dogma was the "savage" stereotype, which ordained that Indigenous peoples "didn't and couldn't own land and form legitimate governments."[63] The *terra nullius* doctrine erased people, culture, history, and relationships among humans, ancestors, and other-than-human beings, while also denying the intelligence and awareness of land. It defined land precisely as rape culture defines women—as innately violable, as "things to be taken."[64] *Possess* means both to own as property as well as for a man to "gain sexual possession of (a woman)" (*OED*). The verb *rape* shares these connotations, meaning both "the seizure of property by violent means" and "the act or crime, committed by a man, of forcing a woman to have sexual intercourse with him against her will" (*OED*).

A poster, "TERRA NULLIUS IS RAPE CULTURE," showing two drilling oil rigs, is a collaboration by Iako'tsi:rareh Amanda S. Lickers (Seneca, Turtle Clan) and Lindsay Nixon (Nehiyaw/Anishinaabe).[65] The verb *drill* means to "make or bore a hole" (*OED*). It also doubles as a slang term for the action of a man sexually penetrating a woman in a hard, painful way.[66] This is commonly known, even if rarely openly reckoned with. One of the original images used in the 1970s by a feminist antipornography group was an illustration from the satirical magazine *Slam*, which showed a heavily muscled White man in a hard hat standing in front of a brick wall. He inserts a drill that is longer than his leg into the vulva of a naked White woman lying on the ground, with her legs in the air. The tagline reads: "At last, a simple cure for frigidity."[67]

The chant "Drill, baby, drill" was repeated rapturously at the 2008 Republican convention and then taken up by Sarah Palin, the vice-presidential candidate. At face value, the chant urged the Federal government to open up previously protected lands in the Arctic to extraction. Subsequent commercial items featuring Palin—from stickers and T-shirts to a pornographic film—highlighted its sexual implications.[68] (Figure 4.2). Under the Trump administration, the protected lands in question have been opened to oil extraction.[69]

Nuke is another ecologically meaningful word that works as a synonym for *fuck*. For example, one counterprotester at the anti-nuclear Seneca Women's Peace encampment wore a T-shirt with the phrase "Nuke the bitches."[70] Long before *drill* or *nuke*, though, there was *plow* (*plough*), also meaning a possessive and aggressive penetration of a woman by a man. This use appears in a 1664 text, quoted in the *Oxford English Dictionary*: "Is't not a sad sight to see a rich young Beauty . . . subject to some rough rude Fellow, that ploughs her, and esteems and uses her as a chattel?"[71] A contemporary use appears in an online forum, where a 2019 contributor advised older men how to "vag-plow" young women.[72]

Societies that used plows instead of hoes and digging sticks have historically tended to hold the most patriarchal beliefs about women, including that they should be confined to the home under the rule of men.[73] One explanation is that working the plow required greater upper-body strength, leading to a gendered division of labor that valued men's work more highly. The plow's violation of the soil and the plower's concomitant tethering, whipping, and domination of the animals who pulled the device, produced a field of meanings and consequences. Plowing encouraged men to imagine themselves as masters over soil, other animals, and women. The modern steel plow has led to massive erosion and habitat destruction and is a major driver of the Anthropocene.[74]

FIGURE 4.2 "Drill Baby Drill." © Threat Pit Inc.

Midwife, activist, and teacher Tekatsi' tsiah:khwa Katsi Cook (Wolf Clan Mohawk) asks all to realize that for all human beings: "Women are the first environment . . . From the bodies of women flow the relationship of those generations both to society and to the natural world. In this way is the earth our mother, the old people said. In this way, we as women are earth."[75] As that inner "first environment" is defined and treated by any society, Cook suggests, so, too, is the outside one. Ecofeminist thinker Barbara Mor offers similar perception: "For if the female body can be controlled or used, in any way, from the outside—via exploitative definitions or systems— then so, it follows, can everything else."[76] A heteropatriarchal society's definition and treatment of the mother and all it classifies as feminine is the paradigm for its treatment of land. Plow culture, nuke culture, drill culture—all are manifestations of motherfucker culture.[77]

THINKING LAND AND BODY

The doctrine of *terra nullius* encapsulates *Man's* denial of what Vanessa Watts calls "Place-Thought . . . the premise that land is alive and thinking and that humans and non-humans derive agency through the extensions of those thoughts."[78] Returning to the "TERRA NULLIUS IS RAPE CULTURE" poster, we find a similar awareness. Ritually stitched in red thread, on the horizon are several "skydomes," which, to Haudenosee, represent "all life forms who live in equality under the dome of the universe."[79] The skydome presence reminds viewers of "the unseen worlds whose fates affect us, of the potential of the unknown, the power of curiosity, and of Sky Woman herself."[80]

Originally from another world, Sky Woman is the being who becomes land, *terra*, Earth, in the histories of several Indigenous Nations, as explained by Watts:

According to Haudenosaunee, Sky Woman fell from a hole in the sky . . . On her descent, Sky Woman fell through the clouds and air towards water below. During her descent, birds could see this falling creature and saw she could not fly. They came to her and helped to lower her slowly to waters beneath her. The birds told Turtle that she must need a place to land, as she possessed no water legs. Turtle rose up, breaking through the surface so that Sky Woman could land on Turtle's back. Once landed, Sky Woman and Turtle began to form the earth, the land becoming an extension of their bodies.[81]

Watts further explains that land thus is "the literal embodiment of the feminine, of First Woman . . . When Sky Woman falls from the sky and lies on the back of a turtle, she is not only able to create land but becomes territory itself. Therefore, Place-Thought is an extension of her circumstance, desire, and communication with the water and animals—her agency."[82] Hence, the name Turtle Island for North America.

This story of Sky Woman, Watts makes clear, is not myth or legend, but "what happened."[83] Colonialism continually seeks, she says, "to make Indigenous peoples stand in disbelief of themselves and their histories,"[84] but the truth remains that land *is* alive, feminine, intelligent, and able to exert power. Robin Wall Kimmerer also writes of Sky Woman and notes that in the "public arena" her story often is "told as a bauble of colorful 'folklore.' But, even when it is misunderstood, there is power in the telling," for it carries "instructions,"[85] including about the need for respect for the feminine and for the living, aware land. Western thought dismisses the recognition of thinking land as a false belief, a projection of human traits onto the earth. This is a reversal. Without thinking land, Watts and Kimmerer affirm, there would be no thinking humans. *Terra nullius* is the fiction, the lie.

The Western worldview of mindless and rapeable land is mapped in an ad for audio speakers with the tagline "Feel the raw, naked power of the road"[86] The image is of a sandy-skinned, curvaceous, young, and desirable woman. She is shown lying naked on a flat yellow metallic plate. The ad is aimed at a mainstream audience, so the woman lies on her side, with her genitals hidden. Her black, wavy hair fans out around her head; her eyes are closed, and her arms crossed, corpse-like, in front of her chest. Inscribed over her entire body are road maps, showing settler highways, towns, and cities. Indigenous place names, as usual, have been erased.[87]

Anthropocene culture is geo-pornographic, even geo-necrophilic. The more inert and empty the projection of woman-land, the more desirable. In this ad, there is no living Turtle supporting the woman, only a garish metal plate. There are no birds or other beings. Relationships are gone, replaced by roadways. This is a *terra nullius*–type fantasy of women and land without power, minds, and drive. The message is clear: It is only *Man* who thinks, moves, and drives over her dead body.

The words in the copy—*naked* and *raw*—suggest sex and pain. The display of an eroticized feminine corpse is a standard misogynist trope in film and television;[88] it also points to a real-world complex of pain, sex, speed, and destruction. Highways provide well-known dumping grounds for the bodies of sexual-murder victims, most often sex workers, so much so that the FBI established a Highway Serial Killing Initiative.[89] In Canada, there is the terrible "Highway of Tears," an area marked by multiple disappearances and murders of Indigenous women, youth, queer, and two spirit people. Native women in Canada and the United States are trafficked, sexually assaulted, and murdered at the highest rates, mostly by White men, who often face few consequences.[90] In 2019, a report from the Canadian government affirms a prior Native assessment of this atrocity, calling these murders a form of "genocide" for which Canada is responsible.[91] Chief Commissioner Marion Buller further clarified that this is a "deliberate race, identity, and gender-based genocide."[92]

Psychiatrist James Gilligan diagnoses the violence of Western civilization as inseparable from its heteropatriarchal, dichotomously gendered, racist, and economically inequitable character. Women and men are said to be opposite and unequal; men become "violence objects," while women become sex objects. This system, Gilligan observes, institutes a disastrous "code of honor and shame [that] generates and obligates male violence."[93] Masculinity and heterosexuality are never fully assured, but must be proven over and over again. Violence, coded affirmatively masculine, is a most effective way to do that.

Writer Tomson Highway (Cree, Barren Lands First Nation) explains that this heteropatriarchal masculinity was all too apparent in an atrocity committed by four "young, white, heterosexual men against a 17-year old Cree girl," Helen Betty Osborne. The men abducted her, "dragged her off into the forest, and raped her by ramming a screwdriver fifty-six times up her vagina and left her there to bleed to death." Some people, he says, more or less excused this as " 'normal' male behavior." One of the murderers was eventually convicted, but not until fifteen years later. Highway, who "thank[s] his lucky stars every day of my life that I'm not a heterosexual man," attributes this horrific violence to such men's mistaken belief that they are "100 percent male": That's "when the trouble starts. That simply doesn't exist biologically, spiritually, psychologically. We all have elements of both sexes in us. And so men who are scared of that, who are terrified, and who are trying to prove that they are 100 percent male, those are the men that go around ramming screwdrivers up people's vaginas. They're real men."[94]

The report *Violence on the Land, Violence on Our Bodies: Building an Indigenous Response to Environmental Violence* includes a vital toolkit and is available online.[95] This collaborative project from the Native Youth Sexual Health Network and the Women's Earth Alliance compiles years of research, including input from activists associated with Indigenous rights, environmental justice, and feminism. At its core is the idea of *environmental violence*, a concept—generated by elders, wisdom keepers, and activists—that demands attention to gendered and sexualized violence, as well as the classed and racialized violence recognized by environmental justice. *Environmental violence* describes "the disproportionate and often devastating impacts that the conscious and deliberate proliferation of environmental toxins and industrial development (including extraction, production, export and release) have on Indigenous women, children and future generations, without regard from States or corporations for their severe and ongoing harms."[96]

The report includes an illustration by intern Katie Douglas, titled "Violence on our Earth." The drawing shows the land as a naked woman, flat on her back and in obvious pain. An open-pit copper mine has been dug into her womb. One of her legs is missing, replaced by a spewing oil well. A flag is stuck into her other leg. Her right breast is now the stump of a chain-sawed tree, her left a trash can.[97] Douglas says that she intends her artwork to signify the precariousness of a future when there is such "widespread domination of the Earth, Indigenous peoples and women."[98]

That "domination," in heteropatriarchal constructions, MacKinnon theorizes, is synonymous with *sex* and normative gender: "All the ways in which women are suppressed and subjected—restricted, intruded on, violated, objectified—are recognized as what sex is for women and as the meaning and content of femininity."[99] So too, violence becomes the meaning and content of masculinity: "Aggression against those with less power is experienced as sexual pleasure, an entitlement of masculinity."[100] The basic heteropatriarchal order can be expressed in one succinct diagrammed sentence, MacKinnon avers: "Man fucks woman; subject, verb, object."[101]

American Studies scholar Rina Swentzell (Kha'p'oo Owinge, Santa Clara Pueblo) gives a similar sexually political analysis, calling out the atrocious practice of "power as an integral part of sexuality . . . That is what the Inquisition was all about. That is what the whole conquest of the Southwest was about—power and control by males."[102] That rapist conquest is then portrayed in standard American myth-making as heroic and manly frontier crossing.

In Toni Morrison's classic *Beloved*, Baby Suggs, the unchurched preacher and great-hearted holy woman, disputes this creed and offers wisdom about the need to respect limits: "Everything depends on knowing how much . . . Good is knowing when to stop."[103] Some "ecomodernist" advocates of technological-capitalist expansion challenge this wisdom, such as Ted Nordhaus and Michael Schellenberger, who champion a slogan of "Break Through." They chide green thinkers and activists for speaking of "ecological problems as the inevitable consequence of humans violating nature." This, they argue, constructs a binary in which human activity is seen as an "intrusion on . . . a separate and once pure nature."[104] This argument does not hold though. Numerous feminist and/or green thinkers and activists always have insisted that humans are part of nature and, moreover, that it is not all humans who are causing the global ecocidal problem, but only those who hoard power and goods. Nordhaus and Schellenberger go on to urge environmentalists to be more positive, to stress prosperity as a goal, and no longer rely on verbs like " 'stop,' 'restrict,' 'reverse,' 'prevent,' 'regulate,' and 'constrain.' "[105] While equal access to a community's "natural common goods" (Cáceres) and shared prosperity is a fine ideal, this requires words like "stop" and "constrain" to address routine frontier violation, extraction and theft, occupation, trashing and dumping. Respect for the word *stop* is so necessary because the environmentally abusive culture is at root a rapist one.[106]

"FRACK OFF"

The website for Independence Institute.Org profiles "free market energy policy" advocate Amy Oliver Cooke, reporting that she favors the slogan "Mothers in love with fracking."[107] When Cooke was appointed by President Trump to an

Environmental Protection Agency team, one green organization, Roll Forward Environment (RFE), showed a picture of
Cooke wearing a T-shirt sporting that slogan and stamped the
words "Mother Fracker" over the photo. Activists regularly
come up with related lines, including "Frack off," "Stop the
mother-fracking," "No permission asked, no permission given.
Fracking is rape,"[108] and "Don't frack your Mother" (Figure
4.3). All of these play on the resonance of *frack* and *fuck* and
implicitly recognize the similarities between *fuck*, with its gendered fusion of sex-and-violence, and the mechanics and logic
of the fracking process itself.

Hydraulic fracking starts with drilling, enhanced by pressur-
ized tons of water treated with noxious chemicals that enable the
drill to penetrate deeply into rock formations. In the case of oil
extraction, tons of waste, much of it highly radioactive (barium
and radium) is produced. The fracking process continues as coarse
materials like sand and ceramic are forced into the newly drilled
holes to "coerce the clefts into remaining open, thereby allowing
continuous extraction of energy, gas or oil."[109] The similarity to
rape includes the invasive drill, the extraction of energy, and the
involvement of poisonous waste.

If this were not a rapist culture, hydraulic fracking might
not even occur to anyone, since everyone—not just those in
the directly affected communities—would understand it as a
violation. Fracking's hostile invasion of the ground to extract
power and the concomitant poisoning of water would not make
sense, might even be recognized as "psychopathic" (Bambara).
People, instead, would create societies that take into consideration human as well as other-than-human beings, societies
that *respect* elemental power, knowing that without this respect
catastrophe will, sooner or later, follow.[110] If this were not a
motherfucker culture, there would be well compensated and
meaningful jobs that don't rape the Earth. There would be fuel,
food, technological aids, medicine, and other necessities and
joys of life, but these would be based in compassion, respect,

FIGURE 4.3 Michelle Sayles, "Don't Frack Your Mother." Berks Gas Truth Logo.

equity, reciprocity, sensible and agreed upon limits, and an embrace of "human simplicity."[111]

To further understand the relationship of fracking and rape culture, it helps to consult Teresa Brennan's theory of the "transmission of affect." Brennan begins with this basic precept: "All beings, all entities in and of the natural world, all forces, whether naturally or artificially forged, are connected energetically."[112] Everything we experience involves these energies and their exchanges—which become inequitable and exploitative under conditions of oppression. This destructive transmission of affect is intrinsic to sexual violence.[113] The sexual abuser (characteristically a man in heteropatriarchy, but women can do this too) enhances himself by tapping into, taking, and consuming the life force of his victims, leaving them depleted. He simultaneously relieves himself of his psychical wastes or negative affects—fear, shame, anger, and guilt—by transferring them to his victims, who experience these terrible feelings as their own. Both extraction and dumping are as intrinsic to sexual abuse as they are to ecological abuse.

This pattern is observed, regularly, by survivors of sexual assault. In *After Silence: Rape and My Journey Back*, Nancy Venable Raine recounts her awareness that her life energies were being actively devoured by the man raping her in a depleting "energy exchange."[114] Though she survived, she also found that the other side of that exchange was that she had been filled with *his* anger, hatred, shame, and guilt. She healed in part by reconnecting with what she calls the "mysterious life-affirming force"[115] found in other-than-human nature.

In their book on the Capitalocene, Moore and Patel identify a rapacious capitalism's need to constantly develop new frontiers for extraction, exploitation, and dumping of waste.[116] These frontiers can be external or they can be internal, like the DNA of Indigenous peoples or the sexual integrity of another person.[117] One incest survivor, Judith Bierman, describes the cumulative horror of childhood rape by a trusted and loved relative, including the experience of having her body used by her rapist as "a last frontier"—being opened and then "mined," when that rapist also built a fence around her to keep others out and "to mark his territory."[118] Another survivor, Marty O. Dyke, denounces her incestuous father as "the rot that festers this Earth," polluting the planet, including the planet's "psyche" and hence all of "our lives."[119] The rape of bodies and the rape of land extract energy while dumping toxic wastes. This is a physical and psychical contamination, resulting, if the being who has been raped is left without access to healing, in disintegration.

Those who have been sexually assaulted speak of trauma that is "etched" into to their "souls."[120] Tarana Burke, a survivor of three assaults, is the founder of The "me too" Movement™ and Just BE Inc., which is "focused on the health, well-being and wholeness of brown girls everywhere!" Burke writes that this work "started in the deepest, darkest place in my soul."[121] Sarah Deer concurs that "worldwide, victims of rape often describe their deep, psychic wounds as harming the very foundation of their identity . . . something critical to our being and personhood."[122]

In parallel ways, egregiously poisoned land becomes a "national sacrifice zone: an area so contaminated or so depleted of its natural resources that it is unlikely to be able to sustain life."[123] The people who belong to these lands suffer similar spirit or soul violations, manifesting in depression, anxiety, and increased rates of violence and victimization, including suicide.[124] The concept of "land trauma," as developed by Indigenous thinkers and activists, speaks directly to "the emotional and spiritual experiences of loss of land and identity" and can refer as well to "feelings of grief and pain that have been inferred or absorbed through Indigenous lands and waters."[125] Harm to land is harm to the people connected to that land. People feel the rage, grief, and pain of land. So too, the land feels the people's rage, grief and pain.

CHATTEL SLAVERY: THE SOURCE
MADE RESOURCE

The exhibition *White Shoes* is a set of ten searing photographs by Nona Faustine. The artist poses in New York metropolitan locations with specific meanings to the legacy of chattel slavery, essentially, she says, making "memorials" with her body.[126] In each photograph, she appears mostly naked, though always wearing white high-heels, the shoes emblematizing White supremacist-patriarchy and the sexualized destructiveness it has meted out to Black women.

New York State allowed slavery until 1827. One of Faustine's indelible photographs, made in 2013, is set on Wall Street, now a metonymy for global capitalism, on the spot of a former slave market. Faustine stands atop an auction-block-like box, staring straight ahead, with her hands clasped in front of her belly. The photograph is titled *From Her Body Sprang Their Greatest Wealth, Wall Street* (Figure 4.4).

One journalist, appreciating the series, perceives its message to be that slavery "reduced people to mere bodies, machines of muscle."[127] This is indisputable; slavery was torturous and atrocious for both men and women. But Faustine's *Wall Street* speaks to chattel slavery's use of women's bodies. Harriet Jacobs, in *Incidents in the Life of a Slave Girl*, was one of the first to tell this truth: "Slavery is terrible for men; but it is far more terrible for women. Superadded to the burden common to all, they have wrongs, and sufferings, and mortifications peculiarly their own."[128] White masters used Black women as muscle for sure, but also uniquely as women by seizing, exploiting, and commodifying their sexual, procreative, and nurturant powers.

Writing in 1983, historian and civil rights activist-scholar Barbara Omolade perceived that White men "would continue to penetrate and plunder Mother Africa," while building a society over the forced production and reproduction of black women.[129] Enslaved Black women, Omolade recognizes, were fragmented

FIGURE 4.4 Nona Faustine, *From Her Body Sprang Their Greatest Wealth, Wall Street*. From the *White Shoes* series (2013). Archival pigment print, 30″ × 40″, 76.2 × 101.6 cm. ©Nona Faustine.

into body parts: "Her vagina, used for his sexual pleasure, was the gateway to the womb, which was his place of capital investment— the capital investment being the sex act and the resulting child the accumulated surplus, worth money on the slave market."[130]

Poet M. NourbeSe Philip bitterly speaks of the ways that the colonial-slaver's "royal will and pleasure" decreed "that the space lying between the legs of the black female is hereby declared a thoroughfare. A black hole."[131] In her inspired essay/poem/play, "Dis Place—The Space Between," she keenly analyzes the terrible parallels between the enslaver's oppression of this "space" and the establishment of the plantation system:

> Whether we conceive of the space between the legs as one space, the cunt; two spaces, the cunt *and* womb; or one continuous space extending from cunt to womb, control of and over this space or spaces is a significant marker of the outer space . . . The

space between the Black woman's legs becomes. *The place.* Site of oppression—vital to the cultivation and continuation of the outer space in a designated form—the plantation machine.[132]

Philip notes that she uses the word *cunt* "advisedly" to explicitly reject patriarchal language that either medicalizes or pornographizes the female sex as a way to separate "us from our bodies."[133] The charged word *cunt* also helps to illuminate the ways that the geo-pornographic patriarchal culture acts to own and control the sexual-reproductive inner space of women:

> The Body. And that most precious of resources—the space. Between. The legs. The Black woman comes to the New World with only the body. And the space between. The European buys her not only for her strength, but also to service the Black man sexually—to keep him calm. And to produce new chattels— units of production—for the plantation machine."[134]

Philip concludes that the "inner space between the legs" of the Black woman would become the very "fulcrum of the New World plantation."[135]

In 2016, Ned and Constance Sublette published *The American Slave Coast: A History of the Slave-Breeding Industry.* This is an excellent history, albeit one that does not acknowledge most relevant prior Black feminist analyses, including those cited here (Barbara Omolade, M. NourbeSe Philip, Adrienne Davis). The Sublettes detail "a history of the making of the United States as seen from the point of view of the slave trade."[136] This trade, they conclude, depended upon Euro-American men's use of the Black woman as a "capitalized womb . . . the essential production engine of the slave-breeding economy, which in turn fueled a global economy that processed slave-grown cotton into mass-produced cloth."[137]

The history of the rape-based slave-breeding industry, the Sublettes say, is simultaneously a "history of money."[138] For wealthy Southerners in the United States, slaves *were* money. Black "people were capital, children were interest, and women were routinely violated."[139] Twelve of the first fifteen presidents were legal owners of other human beings. Fourteen-year old Sally Hemings, enslaved at Monticello, was taken by forty-four year old Thomas Jefferson to France to serve his daughter. During that time, Jefferson raped Hemings and established a continuing sexual possession of the girl born in slavery, resulting in at least six children after they returned to Virginia. An exhibit at Monticello (installed only in 2018) uses the word "rape" with a question mark[140] because of that continuing relationship, but that additional punctuation is misguided. The social power differential between the two could not have been more drastic, negating Hemings's ability to consent in any full sense. Using the word *rape* does not deny her agency, within that context, to try to get the best situation for herself and her children. Rather, it acknowledges her humanity and Jefferson's culpability.

Jefferson, Andrew Jackson, and James K. Polk, like "all slaveholders," kept their wealth "primarily stored in the form of captive human beings."[141] They used their political power to push for the expansion of slavery into new regions, seizing more Native land and continuing the genocidal removal of Native peoples to increase their personal wealth.

The African slave trade had ended in the United States by 1808. Jefferson and other men of his class framed this as a humanitarian event. It was not. Rather, stopping the importation of enslaved peoples was another way for the enslavers to make more money in their own slave-breeding industry. This was fully in keeping with capitalism's incessant need for growth—growth, the Sublettes write, "tied directly to the productivity of the capitalized womb."[142] Nona Faustine's *White Shoes* project is remarkable indeed as she ties this founding institutionalized rape of Black women not only to the success of the slave market, but also to "The Market," epitomized by

Wall Street. Faustine recognizes plantation slavery's paradig-matic motherfucking as the fulcrum of global capitalism and thus also the Anthropocene.

Anthropologist Anna Tsing writes insightfully about the Anthropocene as a *Plantationocene*. She defines plantations as "those ecological simplifications in which living things are trans-formed into resources—future assets—by removing them from their lifeworlds. Plantations are machines of replication, ecolo-gies devoted to the production of the same."[143] Tsing points to the ways that plantations destroy land's integrity, replacing relation-ships among diverse life forms with controlled monocultures. Her analysis could be deepened by considering the ways that the plantation slave masters' use and abuse of land paralleled his use and abuse of enslaved Black women, making them the "fulcrum" (Philip) for the "plantation machine"—their reproductive powers providing the exploited "resource," their children counted as the masters' "future assets."

Though she does not directly address sexual assault, Tsing does invoke it in her title—"Earth Stalked by Man." Stalking is a pattern of consistent and often inescapable harassment, commit-ted by both men and women, but mostly by men. Male stalkers are the most dangerous, as many victims of femicide or feminicide[144] (gender-based murder of women by men) first are stalked by their murderers.[145] The reach of the femicidal stalker has been greatly expanded through technologies, notably those of surveillance. So too has the reach and damage of the ecocidal stalker (plowing, nuking, fracking). Tsing is right that in the Anthropocene, *Man* is stalking (Mother) Earth.

Historian Edward Baptist, in *The Half Has Never Been Told: Slavery and the Making of American Capitalism,* devel-ops a sexually political analysis, though he does not credit the directly relevant feminist theoretical-activist work—for exam-ple, Andrea Dworkin. Catharine Mackinnon, and Barbara Omolade—that preceded his. Baptist recognizes the opera-tion of a "white men's code of masculinity" shaping plantation capitalism,[146] a code that boiled down to White men being able to "fuck" at will:

> The tree-turned-into-a bush [cotton] . . . is fucked. So, too, is the soil. When the enslaved men broke it open for the entrepreneur, he fucked this dirt with them as his tool. He fucked this field. He might fuck their wives out in the woods, or in the corn, when it is high. Or their daughter in the kitchen, then the next new girl he buys at New Orleans . . . he fucks the men too . . . Plants, ecosystems, people strain to live their lives according to their own codes, but he twists their efforts into helixes of his own design . . . marking them and their world with his desire."[147]

The Man's possession and exploitation—of girls, boys, women, enslaved men, cotton, corn, dirt, land—are indeed ways of "fucking," more precisely *motherfucking*, life. This assaultive pattern continues to drive such new modes as synthetic biology, when it takes up that professed goal of rewriting the codes of life, disintegrating the core or essence of other beings and replacing these with the scientists' designs and desires.

In sum, the rapist removal and genocide of Indigenous peoples and theft of their land made possible the plantation, which ran on elite White men's rape and capitalization of the Black woman's sexual, reproductive, and nurturant powers. All this then made possible the global cotton industry and globalization itself. Historian Sven Beckert summarizes the conclusions he, Baptist, and other analysts of global capitalism have drawn regarding the significance of plantation chattel slavery and its main crop of cotton: "We cannot know if the cotton industry was the only possible way into the modern industrial world, but we do know that it was the path to global capitalism."[148] This path is a thoroughfare paved over women and others' raped, murdered, and dumped bodies, the same road that drives straight into the Anthropocene-Capitalocene-Plantationocene. It is a Highway of Tears, a road made and mapped by motherfuckers.

COLOR MOTHER NATURE GONE

MUSICIAN TOM BEE'S (DAKOTA) SONG "Color Nature Gone" (1973) speaks directly to the matter of Mother Nature-Earth in the Anthropocene:

> *The color blue no longer paints the sky*
> *And the rivers have turned to grey.*
> *The color green no longer paints the leaves.*
> *And the trees have all burned away.*
> *People have made the colors change.*
> *Color nature gone...*[1]

Color here is the active force. Colors are the *painters*, and the painting—usually—is ongoing. But people have behaved in ways that have caused the familiar hues to take their leave. Bee mourns this *colorlessness* as indicative of the loss of (Mother) nature. There remains only an unnerving likeness—a colorless, un-regenerative, and dis-spirited Stepford substitute, like a grey river to replace what has gone.

La Loba Loca, a "brown ecofeminista" activist and practitioner of "abuelita knowledge"[2] (knowledge passed-down from the grandmothers), is similarly concerned about the departure of color. She condemns a veritable motherfucking as its cause, specifying the production of cocaine (nicknamed "snow" because of its whiteness): "The coca leaf is considered the *Mamita Cuca*, the Mother *Cuca*, a very nutritious, beautiful plant that has been completely raped . . . [in] a completely unconsensual process," to

Call Your "Mutha'". Jane Caputi, Oxford University Press (2020). © Oxford University Press. DOI: 10.1093/oso/9780190902704.001.0001

produce cocaine, the substance that "creates so much blood and addiction and death." La Loba Loca recalls that in its integrity, "the coca leaf is beautiful and it is medicine and it is green and not white."[3]

Color is presence. Color is motile. Color wheels. Color expresses being, speaks, attracts, emotes, and delights. Whatever one's visual capacities, color matters. The blind and deaf Helen Keller evoked color in her writings, as in her phrase "a mist of green." For this, her critics charged her with "chicanery." Keller responded that even though she had no "tactual evidence" of light and color, she believed that "all humanly knowable truth is open to" her and dismissed her accusers as "spirit-vandals."[4]

Keller affirmed the teaching of Annie Sullivan, holding that "the best and most beautiful things in the world cannot be seen, nor even touched, but just felt with the heart."[5] Native scientist Lame Deer (Lakota) similarly instructs on the heart's perceptual powers:

> We Indians live in a world of symbols and images where the spiritual and the commonplace are one. To you symbols are just words, spoken or written in a book. To us they are part of nature, part of ourselves—the earth, the sun, the wind and the rain, stones, trees, animals, even little insects like ants and grasshoppers. We try to understand them not with the head but with the heart, and we need no more than a hint to give us the meaning.[6]

Heart knowledge is imperative to manifesting an alternative world to the Anthropocene. Activist and artist Nigit'stil Norbert (Dene), who describes herself as a "protector" and a "defender of Mother Earth," incorporates heart symbolism in her work because, she explains, the "human heart" has an "intrinsic connection and relationship to the health of our water and land."[7] It is through the heart that we feel-know our most intimate identity with, water and land.

Consider the blueness of sky and sea. Something new, we know, comes "out of the blue." This is more than a figure of speech, for blue is "the deepest color," the place where the real and the imaginary meet, and one place where the "infinity" Hurston speaks of can be experienced.[8] Greenness, too, is instructive. Green appears in the bud of spring and the rot of decay, revealing the twinning of life-and-death in nature's cycle. Greenness is an engineer of photosynthesis, giving food as well as breath, the physical counterpart of spirit.[9] When the integrity of greenness is disrespected and greenness goes away the effect is, truly, as if all the oxygen-spirit has left the room.

Without the colors that so generously paint the Earth there would be no life as we know it. Some of these appeared about 3.8 billion years ago when ancient cyanobacteria began to photosynthesize, producing oxygen and carbohydrates. Primal "photosynthetic pigments impart a rainbow of possible colors: yellow, red, violet, green, deep blue, and blue-green cyanobacteria are known."[10] It was green chloroplasts in plant life who so notably evolved in relationship with cyanobacteria. The colors that paint nature continuously code how sunshine, leaves, birds, and flowers (among others) communicate with one another.[11]

Tomson Highway perceives political-spiritual as well as sexual and gendered meanings of color, bemoaning the dispirited "black and white world of he and she" and yearning for all the living "colours . . . the other genders" that make up life while also making life "worth living."[12] A similar spectrum is invoked by scholar Omise'eke Natasha Tinsley as she hails Black queer genders, while recalling "Danbala, the simultaneously male and female rainbow serpent" of Vodou.[13] This colorful reality is emblematized in the rainbow flag, the original lesbian and gay pride symbol designed by drag queen Gilbert Baker in 1977. The rainbow flag appears continuously in new iterations, reflecting the expanding universe of non-binary, queer, and trans being.[14] Activists of color suggest adding black and brown stripes to the flag to acknowledge and undo racism in the LGBTQ community.[15]

Artist and leader of Black women's revolutionary art Betye Saar connects color to the presence of nature. She salvages, collecting appealing items, from feathers to tin cans, which she groups by hue, explaining: "You can't beat Nature for color. She's got it down."[16] The spiritual meanings of color, as well as color's connection to Mother Nature-Earth, are addressed by Paula Gunn Allen in *The Sacred Hoop: Recovering the Feminine in American Indian Traditions*. As she tells it, the life force-source *is* spirit, *is* color, is thought, and is "fundamentally female":

> There is a spirit that pervades everything, that is capable of powerful song and radiant movement, and that moves in and out of the mind. The colors of this spirit are multitudinous, a glowing, pulsing rainbow . . . she is the true creatrix for she is thought itself, from which all else is born. She is the necessary precondition for material creation, and she, like all of her creation, is fundamentally female—potential and primary.[17]

A society in keeping with this spirit accords honor, Allen continues, to a "diversity of people, including gay males and lesbians" and two spirit people.[18] Oppositional sex and gender binaries/hierarchies have no hold under this rainbow. As Allen tells it, *everyone* is "female" or "feminine" in this foundational way.[19]

Allen scorns the ill will, ignorance, and misogyny that would "assign to this great being the position of 'fertility goddess' . . . [She] bears, that is true. She also destroys."[20] "Living and dying," as Allen elsewhere puts it, are "twin beings, gifts of our Mother, the Earth."[21] In the Anthropocene, the most powerful seek immortality, while the poorest increasingly experience premature death. Meanwhile, an encroaching colorlessness bespeaks a coming world from which nature is "gone." Tellingly, *Man* claims this colorlessness, representing it as a refined and rational aesthetic, even a fashion statement—first as a signature *basic white* and then, with the invention of artificial color, a "synthetic rainbow,"[22] one fashioned quite consciously to "beat" or outdo Mother Nature.

BASIC WHITE

Thinkers like Toni Morrison and L. H. Stallings uphold the sensual-sexual-spiritual glory of "funk," a force saturating the pulsing rainbow with strong aromas, intense flavors, and peak pleasures.[23] In Morrison's *The Bluest Eye*, a plastic baby doll with cold blue eyes, pink skin, and blond hair becomes a perfection idol for a White supremacist and funk-phobic society. Morrison condemns that system's aim to "get rid of the funkiness. The dreadful funkiness of passion, the funkiness of nature, the funkiness of the wide range of human emotions."[24] The tragic Pauline Breedlove (before being broken in and by racism) had experienced the intense funky pleasure of orgasm as akin to the pulsing rainbow:

> little bits of color floating up into me—deep in me. That streak of green from the june-bug light, the purple from the berries trickling along my thighs, Mama's lemonade yellow runs sweet in me. Then I feel like I'm laughing between my legs, and the laughing gets all mixed up with the colors, and I'm afraid I'll come, and afraid I won't. But I know I will. And I do. And it be rainbow all inside.[25]

When the rainbow funk is obliterated, the colorlessness of *basic white* takes over. This is not the color *white* that paints snow, flowers, clouds, bones, seashells, milk, and my hair in later life. Nor is it the color hailed by Leonardo da Vinci as the white that "may be said to represent light without which no colour can be seen."[26] Rather, *basic white* is the purportedly "pure whiteness"[27] invented by a racist society to set in opposition to funkiness and "blackness." The latter, Delores Williams writes, is coded as "frightening, dangerous, repulsive, and a prime candidate for destruction."[28] When *Time* magazine devoted an issue to the nature of evil (June 10, 1991), the cover was entirely, absolutely black. Such color-coding sets up a world where "nothing black can be violated, because illegality and disaster are associated with blackness."[29]

Basic "white space"[30] announces and enforces the elevation of the master class over and separation from "black space" (places where Black people are) as well as spaces defined into nature. Its essence is boiled down by W. E. B. DuBois: "Whiteness is the ownership of the earth forever and ever. Amen!"[31] In her striking poem "White Things," (1923) written after the lynching of a pregnant woman, Anne Spencer (of both African and Seminole descent) takes up much the same theme. She remarks upon the innately "colorful" life-force, so vividly manifesting in "sky, earth, and sea," but "blanched" by an abusive whiteness with its "wand of power."[32] Perhaps Morrison had Spencer's words in mind when, in *Beloved*, the escaped slave Sethe, about to be apprehended, murders her child rather than see her returned to slavery. Sethe's mother-in-law, Baby Suggs, subsequently slips into despair, "roused once in a while by a craving for color and not for another thing," crying out, "Those white things have taken all I had or dreamed . . . and broke my heartstrings too."[33]

This sort of heartbreaking whiteness sets out to stifle heart knowledge and exemplifies European colonialist "chromophobia,"[34] whereby color is associated—negatively—with the feminine, the primitive, and the chaotic. Whiteness, represented as a kind of colorlessness, is identified with superiority, taste, class, and purity. For example, foods like sugar cane, wheat flour, and rice are industrially blanched, stripping them of nutrition but emblematizing a supposed superiority.[35] Whiteness is the sign, moreover, of property,[36] modernity, rationality, cleanliness, civilization, sterility, science, and objectivity. It is taken for granted as the color of the laboratory, the lab coat, the research center, the space suit, and the space ship. It is the iconic color associated with computers, cryogenics, robots, artificial intelligence, the posthuman,[37] the immortal, God, and the future, the latter enshrined indelibly in 1968 in the film *2001: A Space Odyssey*.[38] That revered film actually compares unfavorably to the camp classic *Barbarella*,[39] also from 1968, starring Jane Fonda as a multiply orgasmic, big-haired astronaut steering her own hot-pink spacecraft.

One particularly telling white-is-right future projection is an ad from 1982 for Compuserve, the first significant commercial online service provider.[40] The headline announces: "Welcome to someday." A "clean-cut" White man, dressed all in white, sits at a white desktop computer. A White woman stands alongside him, her hair neatly bound in a bun, wearing all-white clothes, including white high heels. All the items in this controlled indoor space—the chairs, couch, rug, desk, bookshelves, knick-knacks, trash can, and lamp—are white. There is nary a window, a green plant, a swimming goldfish, an orange cat, or a spotted dog. The copy promises that soon (white) people will be able "to shop and bank electronically, read instantly updated newswires" and engage in other conveniences—and all from this purely white space. One might wonder how the place stays so immaculate, whether there is a servant somewhere out of sight, and, if so, what might be their skin tone. Also kept hidden are the dangerous spaces where workers labor to extract the constitutive materials and fuels and the sacrifice zones where the inevitable waste products are dumped. Denial of waste, it seems, is proportionate to a culture's generation of waste.

In *Tar Baby*, Toni Morrison excoriates a White colonizing culture's characteristic movement "to defecate over a whole people and come there to live and defecate some more. . . . That was the sole lesson of their world: how to make waste, how to make machines that made more waste, how to make wasteful products, how to talk waste, how to study waste, how to design waste, how to cure people who were sickened by waste so that they could be well enough to endure it, how to mobilize waste, legalize waste and how to despise the culture that lived in cloth houses and shit on the ground far away from where they ate."[41] Colonizers and technocrats, Morrison is saying, imagine themselves as White—that is, clean, advanced, pure, and progressive, and they view those they subjugate ("shit on") as foul, dirty, smelly, dark, and backward.

The normalized association of whiteness with cleanliness has helped higher status White people in America ensure that their spaces are sanitary and safe. As part of that, governments, corporations, and militaries locate polluting industries, military test sites, and waste-dumping facilities in areas where people of color and poor White people (demeaned as "trash") live and work. This is the essence of environmental injustice.[42] Of course, the Earth is *one* system; toxins circulate and ultimately affect all. Alice Walker breaks it down: "While the Earth is poisoned, everything it supports is poisoned. While the Earth is enslaved, none of us is free. . . . While it is 'treated like dirt,' so are we."[43]

A cover of *Wired* magazine from 2015 highlights a story on utopian future cities, *basic white* ones. The background color behind the image is light concrete grey. In the center is a photo of a sculpture titled *Globopolis*, showing the whole Earth from the vantage point of space.[44] The planet no longer glimmers blue-green, but is now entirely covered with white and impossibly tall skyscrapers. Three words (sounding much like eugenicist propaganda or advertising copy for an AI product) describe this coming paradise: "Beautiful, Smart, Human." Poet Ed Roberson evokes the wonder of a "city life within a living Nature,"[45] but this "Globopolis" instead says that wondrousness is the absence of living Nature. There is no brown dirt, no blue waters, no green grass and plants, and apparently nobody living—human or otherwise. It is a neo-necropolis, uninhabitable for anyone except the lifeless, deathless, colorless, and funk-less from whom nature has gone.

Seven decades ago Rachel Carson diagnosed the disturbances that underlay the Globopolis fantasy and offered at least a partial remedy in the realities of living Nature:

> Mankind has gone very far into an artificial world of his own creation. He has sought to insulate himself, with steel and concrete, from the realities of earth and water. Perhaps he is intoxicated with a sense of his own power, as he goes farther and farther into experiments for the destruction of himself and his world. For this unhappy trend there is no single remedy—no panacea. But

I believe that the more clearly we can focus our attention on the wonders and realities of the universe about us, the less taste we shall have for destruction.[46]

Teresa Brennan also theorizes the reasons that destruction and artifice have become so terribly desirable. She argues that when a commodity culture binds living nature into fixities like plastic, steel, concrete, and consumer goods, the amount of freely circulating energy in the world is reduced.[47] This is because these fixities, particularly plastics, take so long to break down, returning their elements to the life stream to be transformed into new life. Despite the appearance of acceleration, there actually is a slowing as people enmeshed with these fixities become less lively, less colorful, more numbed, and more destructive of living nature, which no longer seems so interesting, worthwhile, or even real.

REMOVING THE "MOUNTAIN OF MANURE"

Civilization's obliteration of living nature, accompanied by a glorification of whiteness, is nothing new. In *Eugenics in the Garden: Transatlantic Architecture and the Crafting of Modernity*, art historian Fabiola López-Durán exposes a clearly basic-whitening and eugenicist agenda at the 1922 world's fair in Rio de Janeiro. The Expoção International do Centenário commemorated the founding of Brazil, its entry into modernity, and a national agenda for "a new race." "World's Fairs," López-Durán explains, have "always been showcases for the products of nations' progress, manifested in grandiose monuments, machines, technological and scientific breakthrough exhibitions, and even in disturbing displays of those 'other' people and cultures to justify racial stereotyping and the discriminatory practice of differentiating between the civilized and the savage."[48] They celebrate and sometimes directly enact *Man's* cleansing campaign against nature, including people identified *as* nature.

Making way for progress was the figurative and actual directive of the 1922 fair. To make space for the show, city authorities
ordered the demolition—really *murder*—of a mountain being in
the center of the city. Mountains are beings who fully manifest the
mothering power. Mountains are alive. Water springs from them.
Trees and other green plants grow on them, filtering air and water
for all, and keeping the ground together with their roots. All sorts
of persons, human and not, live on them and look to them for
wisdom and guidance, and mountains themselves have exceed-
ingly long life spans. Greg Cajete explains that in what is now
North America, where Indigenous people have lived "for probably
30,000 years or more," Tewa elders "remind us of the importance
of the long view when they say, 'Pin peyeh obe'—*look to the mountain*. They use this phrase to remind us that . . . in dealing with
the landscape we must think in terms of a ten thousand- twenty
thousand- or thirty thousand year relationship."[49]

This murder of the mountain being in 1922 meant the
obliteration of knowledge as well as an entire community of
plant, animal, and human beings. Some four thousand people
were immediately made homeless.[50] Originally the home of
the Tupinambá people, Portuguese colonial authorities, two
years after Rio's founding in 1565, named the mountain Morro
do Castelo and established this place as their city center. By
1922, the area had become impoverished, with a population
consisting mostly of people of Indigenous and African descent,
deemed "dirty."[51] Rio city authorities mandated that the mountain be destroyed, they explained, because the entire area had
become an impediment to health as well as the "very negation
of modernity itself, a reservoir of vice and diseases, a place of
a 'marginal' population—mostly poor blacks, prostitutes, and
former slaves—who, according to the elites, invaded the center
of the city 'with their embarrassing practices of superstition
and misery.'"[52] The place was condemned as a "'montanha
de estrume' (a mountain of manure) that, together with its
'undesirable' inhabitants, needed to be removed from the city
centre."[53]

Heteropatriarchal pornography, not surprisingly, makes similar associations. Alice Walker acerbically comments, "Where white women are depicted as human bodies if not beings [in pornography], black women are depicted as shit."[54] A cartoon in *Hustler* depicts a Black man wiping his anus after defecation, turning the cleaned area white, equating dark skin with feces.[55] Scholar Jennifer Nash confirms the prevalence of this racist equation, but argues for complex readings of pornography—as when Black women find pleasure in blackness and anality.[56] In White supremacist projections there is no such nuance, however. President Donald Trump, speaking in favor of migration to the United States from countries like Norway, condemned the entry of people from "shithole countries" like Haiti.[57]

In Brazil, authorities obliterated (cleansed) the "mountain of manure" and the *Expoção International do Centenário* soon was installed, with basic whiteness embedded into its scientific, social, and aesthetic components—including white neocolonial buildings, gleaming mechanical products showcasing progress, and the very hues of the publicity materials. In the latter, "white men, women, and children of classic Greco-Roman appearance, wearing white robes and crowns, are arranged against a transplanted European-looking hill and gardens to frame the architecture and machinery of modern factories."[58] Under the manifest mission of establishing hygiene, morality, and beautification, the catalogue implicitly expressed a national desire to "totally eliminate all traces of African and indigenous components of Brazilian culture," as it explicitly extolled a "new race" triumphant in the "battle between man, mountain, and ocean."[59] "Modernity," here and elsewhere, was "equated with whiteness."[60] This "new race" defined itself as dominant over nature, able to take down mountains, remove "dirty" peoples, and stand tall against any encroachments of the ocean.

Swept under the rug in all this cleansing are truths that certainly the mountain knew: all human beings are equally and necessarily "dirty" or carnal. And manure, like darkness, death, and rot is essential to nature's process of creation and renewal. If shit is out of place—in drinking water or mixed into food—it is dangerous. But, in place, it fertilizes and becomes soil, which becomes food, and so on. The culture that pretends it is above dirt is on track to turn the world into a virtually immortal and toxic "shitpile" of plastic, nuclear, and chemical waste.[61]

Basic white also characterizes *Man's* quest to conquer "dirty" death, distilled into the false belief, Suzanne Kelly writes in *Greening Death*, that the corpse "is an actual pollutant."[62] A sort of anti-corpse is pictured on a cover of *Time* magazine from 2011, which announces, "2045 The Year Man Becomes Immortal*."[63] The asterisk is explained: "If you believe humans and machines will become one." The illustration is a side view of a pallid cyborg head set against a white wall. This head is bodiless, hairless (no animal fur or funk here), and comes equipped with a steel-toned power port in the neck. Embedded just above that port is the only color—a greenish plastic indicator, perpetually "on." In this story of the future, eternal basic-white life will soon be available at the touch of a switch.

The same antagonism to carnality results in the signa-ture basic whiteness of robots, including those who simulate human appearance. Hiroshi Ishiguro, the "'father' and bad boy of Japanese robotics," makes androids, mostly girl and woman-appearing ones, originally based on his five-year old daughter (with his wife going along with that). Ishiguro noticed that some of his male students became obsessed with and sought alone time with the simulation, which gave him pause. He now builds adult figures, with the stated goal of making "good-looking mechanical women." When asked to explain, Ishiguro replied, "A beautiful woman you don't picture going to the restroom."[64] No shit.

THE BLACK MADONNA IN THE WHITE CITY, PART I

Another industrial-age fair worth revisiting is the World's Columbia Exposition in Chicago in 1893. This was "Victorian America's equivalent to the modern-day Olympics and Disney World rolled into one."[65] It showcased White American exceptionalism, while honoring the 400th anniversary of Christopher Columbus's arrival-invasion. The fair is familiar to many who have read Erik Larson's best-seller, *The Devil in the White City*, about the misogynist serial killer H. H. Holmes, dubbed "America's Jack the Ripper." Holmes operated in the area at the same time and took advantage of the fair to lure his victims.[66]

"The White City" refers to the fair's signature locale—its "Court of Honor," representing once again the future ideal city. All of its neoclassical buildings were white, and it had brilliant street lights that made even the night white. The area soon was dubbed the "White City and the name stuck. Its impact was described by one enthusiast as "a little ideal world, a realization of Utopia," one suggesting a time "when all the earth should be as pure, as beautiful, and as joyous as the White City itself."[67]

Historian Robert Rydell writes of "the racist underpinnings of" this "utopian artifact"[68]—by which he means the entire Exposition. People of color were excluded from equitable representation in the planning committees and official exhibits, but included in "so-called living ethnological exhibits of 'primitive' human beings." Leading public intellectual-activists Ida B. Wells and Frederick Douglas jointly authored a pamphlet that spoke to the injustice in the planning of the Exposition and racism in the larger culture.[69]

The White City stood in stark contrast to the fair's "Midway" area, which was produced by the Exposition's Department of Ethnology. The Midway featured delights like Ferris wheels and belly dancing, along with people (African, Chinese, Japanese, and Samoan) installed into living ethnographic displays. The exhibits of American Indian cultures stereotypically depicted them as "savage" and "vanishing," outside the stream of progress and history. The Midway, Rydell writes, "provided visitors with apparent ethnological, scientific sanction for the American image of the non-white world as barbaric and childlike."[70]

Significantly, one Black woman was prominently featured in the White City. Nancy Green, who had been born into slavery, was hired by the Davis Milling Company, which made Aunt Jemima pancake flour, to play the "Mammy" stereotype—the large, grandmotherly Black woman shown as the warm and happy nur- turer of white families. American studies scholar Micki McElya, in *Clinging to Mammy: The Faithful Slave in Twentieth-Century America* observes that the stereotypic "Mammy" was pictured as the very "antithesis of desirable white femininity" to rebut charges from abolitionists and runaway slaves describing White men's "rampant, violent sexuality" along with their "fathering of black women's children." As "Mammy" was said to be "like one of the family," McElya writes, this image served to divert atten- tion from White masters' and overseers' actual kin relationships to the enslaved children whom they had fathered.[71] The "Mammy" stereotype is made to tell the master's lie, covering up the horrific motherfucking so fundamental to American wealth and progress.

Langston Hughes, in "Lament for Dark Peoples," mourns those "caged in the circus of civilization."[72] Nancy Green was not overtly caged, but she played the role she was given in the White City's Hall of Agriculture, in a booth "designed to look like a giant flour bar- rel." There, she "greeted guests and cooked pancakes, all the while singing and telling stories of life on the plantation."[73] The com- pany spun a mythology around "Aunt Jemima," involving a "secret recipe" only she knew. McElya contends that White Americans' hunger was not so much for pancakes, but for the fiction of the "faithful slave,"[74] which absolved their sins, demonstrated that African Americans were not angry over the injustices they experi- enced, and proved that Black women wanted to take care of White families even more than their own. This "clinging to Mammy" ten- dency remains to this day, she continues, in the persistence of the "Mammy" stereotype so damaging to Black women as workers and as mothers and so noticeable in the sale of racist collectibles.[75]

In a provocative interpretation, occasioned by an encounter with one such collectible showing a Black woman with a cauldron, Alice Walker suggests that behind the stereotype of the "Mammy" remains the presence of the Black Madonna, "the Goddess, She who nurtures all."[76] The Black Madonna is something of a univer- sal deity, well known, artist Karla Dickens (Wiradjuri) affirms, as the "patron saint of misfits and the oppressed."[77]

On the surface, the "Mammy" statue of a Black woman serving food emblematizes the racist ideology that Black women are fit only for food preparation. Though this work is made menial in the culture of *The Man*, preparing food is truly the science and art that provides the basis of human family, community, and life. The cauldron signifies nature's ongoing issuance of sustenance *and* knowledge.[78] Walker reads the cauldron as a most "potent symbol, the big black pot, in itself. For isn't the black pot of a black woman's womb the vessel from which white men secretly fear they came, just as the big black hole in the cosmos is the black pot into which they fear they will disappear?"[79] Walker then relates a dream she heard in a radio broadcast, told by Canadian psychoanalyst Marion Woodman, who had heard it from one of her clients. The setting is a large outdoor party with lots of food, drinks, music, and fun. In the midst of the grounds is a "cage" holding "a large black woman. The dreamer goes up to the woman in the cage and says, 'What are you doing in there?' The large woman replies, 'Who do you think is giving the party?'"[80] *Man* parties on, an ungrateful guest pretending to be the provider, while trying to cage the actual provider and giver of gifts—Nature-Earth.

Like actual Black women, Mother Nature-Earth is no "faithful slave." If the Black Madonna stops powering the gathering, ceases to serve up the food and drink, quits singing songs and telling stories, takes back her recipes, and either turns on all the lights or puts them all out, the party truly is over and the neo-necropolis takes us.

THE BLACK MADONNA IN THE WHITE CITY, PART II

In the Compuserve "Welcome to someday" ad, the man and woman do not touch, but she does pat the computer.[81] In such basic-white spaces it's common to find all warmth and sensuality centered on a machine. A suggestive 2013 ad for Bluebeam Software is typical. A White woman, wearing a short-short white onesie, lies on her back on a white mattress. She stretches her naked legs up a white wall and props a computer tablet up against one. A top headline suggests: "Collaborate in bed." A middle one adds, "Anything is possible." Below appears the company's trademarked slogan "No limits."[82]

More than one hundred years earlier, this same entanglement of technology, sexuality, and disdain for limits was embedded in another display in the White City at the Chicago Exposition. The patrician historian Henry Adams was enraptured by the fair and its exhibit of electric engines and wrote about it in "The Dynamo and the Virgin," a chapter in his classic *The Education of Henry Adams*. Adams hails the dynamo—an electrical generator—as "a symbol of infinity,"[83] one he analogizes to the Virgin Mary, the Sex Goddess Venus, and the *Magna Mater* Diana of Ephesus.[84]

The Virgin Mary was early Catholicism's answer to the demand for a goddess figure in the new religion. Although the church incorporated one, they whitened and disempowered her, making her an asexual virgin and a subordinated mother who, Simone de Beauvoir decries, "kneels before her son," freely accepting "her inferiority."[85] Mary in this theology is not divine. She treads on herself in the form of the Earth Serpent, before being finally assumed into the Father's white heaven.[86]

Diana of Ephesus, whom Adams also relates to the Dynamo, is openly an earthy Black Madonna—with black skin and wearing a multi-turreted crown. There are bull testicles all over her torso (later depicted as human breasts) and a variety of animals form the lower part of her body. *Diana*, whose name means "shining one" or "dual-Ana," was sometimes dark and sometimes light, with no moral judgment implied.[87] Her message is that darkness is not the opposite of light, but rather its womb, root, or source. Shining light is the manifest face of darkness. Sometimes Diana was shown emerging from a dark cave and holding a torch,[88] a configuration that may have been the model for the Statue of

Liberty. Black Diana's worship also continued even as the Catholic Church dominated in Europe, her influence felt in hundreds of artworks depicting the Virgin Mary as a Black Madonna.[89]

Adams rightly dismisses the trivializing notion that Venus or Diana was worshipped for a superficial "beauty." Rather, he says, it was "because of her force; she was the animated dynamo; she was reproduction—the greatest and most mysterious of all energies; all she needed was to be fecund."[90] Adams is partly right, but he still reduces the sexual-mothering power to the reductive "fertility goddess" rebuked by Paula Gunn Allen. Adams refuses to recognize that dark is intrinsic to light, death to life, and that, as Suzanne Kelly affirms, death and rot manifest nature's intelligence,[91] guaranteeing ongoing renewal.

The Exposition's Dynamo in the White City represented a promised perpetual light. Adams does acknowledge that there are a "few tons of poor coal hidden in a dirty engine-house carefully kept out of sight" that were necessarily powering that Dynamo.[92] Still, he never factors coal into his vision of the sacred. Rather, he genuflects only to the whitened Virgin-Dynamo as a commodified womb producing electricity. This denial of the entirety of reality is at the root of Anthropocene Man's inability to acknowledge or reckon with disasters resulting from his reckless use of fossil fuels, with coal being the most dangerous of these.

This splitting of the life force-source continues to rule in the Anthropocene, again indicated by the consumer culture's worship of the whitened Virgin in the machine. At the inception of the Great Acceleration, the cultural theorist Marshal McLuhan surveyed popular culture (ads, newspaper headlines, art), which he called "the folklore of industrial man" (and used as the subtitle of his book *Mechanical Bride*). There he found a telling icon of a "mechanical bride," who sometimes shows as a woman composed of fragmented, replaceable parts, sometimes as a feminized, sexualized machine. McLuhan discerned here a "cluster image of sex, technology, and death which constitutes the mystery of the mechanical bride."[93]

The same cluster is everywhere in the continuing folklore of Anthropocene *Man*, such as in the robotic-erotic voices of digital personal assistants and in sex robots marketed mostly to straight men.[94] It's also obvious in what media scholar Sarah Sharma identifies as the "long line of technologies ('Mommies') that have been designed for taking care of a male-dominated world." As Sharma explains, the phrase "'post-mom economy' emerged to capture this particular moment in tech(bro) culture when Uber ('Mommy, drive me'), TaskRabbit ('Mommy, clean my room'), GrubHub ('Mommy, I'm hungry') and LiveBetter ('Mommy, I'm bored') emerged."[95] The people doing this low-paid labor, of course, are the socially marginalized "shit workers," who nonetheless carry out the most essential, if devalued, work of maintaining life. At the same time, the utter necessity of "Mommy" (even while she is demeaned and unpaid) is implicit. Sharma astutely perceives that "building Mommy into our devices reflects a fear of her departure."[96]

The symbol par excellence of the whitened virgin-machine is the *fembot*,[97] the robotic erotic made in supposedly "perfect" form, white-featured, young, slim, desirable, conventionally abled, and feminine. The fembot, by definition, is not fully human, and is regularly cast as a sexual and/or domestic slave. Her whiteface masks the legacy, once again, of chattel slavery.

Consider the image on the April 2015 cover of the personal finance magazine, *Kiplinger's*.[98] The headline invites readers to "Cash in on the future" and displays a fembot shown from the belly up, posed against a sky-blue background, with matching blue eyes and a dreamy if blank expression. The unlined face is of a pretty, young, slim woman, "flesh-toned," though with plastic-white spots showing through. Around the head orbit rows of rectangles, showing hundred dollar bills as well as blurry digitized scenes (outer space, a radio signal meter, the Pentagon). The fembot's torso is partially white plastic and a bit of pink, but also pale blue, suggesting the signature sky blue of the whitened, heavenly Virgin Mary and the euphemizing blue liquids used in tampon advertising as an apotropaic against the carnal, alarming redness of menstruation. Fembots, by nature, are not capable of "bleeding to life"—Ntozake Shange's summation of the menstrual-mothering power.[99]

Though wearing that painted-on whiteface, fembots and all robots, as analyzed by scholar Jerome Gregory Hampton, normalize the paradigm of chattel slavery. Hampton questions why robots are made in the image of humans; his subtitle avers that "yesterday's slave" is being reinvented as "tomorrow's robot."[100] Consider the "Bixby" personal assistant from Samsung. The app has a gender-neutral name, but a feminine voice. The corporate website instructs users to just, "Say a command . . . Whatever you need, Bixby [does] it."[101] Such devices and programs invite the user to take up the detrimental mindset of the slave-master. One pop culture cliché is the nightmare scenario of robots, now indistinguishable from humans, becoming the rulers over humankind. But isn't this actually the nightmare of a slave owner?[102] Hampton foresees a possible future where "the enslavement of humans or human-like machines with Artificial Intelligence (AI) or anything resembling an independent consciousness" will serve as a new class of slaves. This, he argues, will "produce adverse effects among all participants in such an institution."[103]

In the slave era, Katherine McKittrick writes, "female slave bodies were "transformed into profitable sexual and reproductive technologies." Subsequently, they "come to represent 'New World' inventions and are consequently rendered axiomatic public objects."[104] This rendering underlies all manner of sexualized gadgets, tools, and technologies, including those robotic dolls being put to specific sexual use. One enthusiast, AI expert David Levy, argues that because you can completely control the features of your robot, and even change them any time at will, robots "provide all the characteristics we want in a companion/spouse/lover."[105] Once again, who exactly is this *we*? What Levy describes is everything a self-styled master or mistress would want in a sexual slave, albeit one unable to birth children. Legal scholar Henry T. Greely believes it "highly unlikely" that any mechanical device "will be able to replace the womb throughout development," because the organ is just too complex, "mysterious," and resistant to simulation.[106]

If forced fembot maternity is not a possibility, rape and sexual slavery are. The fembot is not programmed to say, "No" or "Stop." The subtextual sexual slave on the cover of *Kiplinger's* opens her plastic arms as if to say, "All this is yours." It's easy to imagine pushing a button and hearing her emit phrases like those found in corporate PR—"No Limits™," "The Nature of What's to Come," "The Miracles of Science™" "We Bring Good Things to Life," "Where the Future Takes Shape," "It's the Real Thing," "GOBOLDLY," and "Lockheed Martin We're Engineering a Better Tomorrow™."[107] All this boastful blather suggests that *Man* has achieved his goal of motherfucking Nature-Earth into total submission: the fembot is the symbol of his supposed successful appropriation of the cosmic womb and consequent ability to artificially reproduce a world made to his order.

THE PLASTICOCENE

Some environmentalists use the name *Plasticocene* instead of Anthropocene to point to that substance's disastrous environmental impact.[108] A new substance, "plastiglomerate"—a stone that contains sediments fused with molten plastic—is now traceable in the geological record and interpreted as showing evidence for the reality of the new era.[109] Microplastics fuse not only to stones, but to everyone—our cells, our fat, our blood, our breast milk. "Flesh and environments alike," several experts write, "are now being reconstituted through the lingering and residual effects of plastics."[110] Again, it is important to state that economic, racial, and other injustices position some people (because of their location, their work, or their potential to become pregnant and produce breast milk)[111] to take in and also transmit more of the toxic stuff, while also lacking access to health care to deal with the consequences.

Plastic, like cocaine, is a refined substance that speaks through color change. Though plastics are now made mostly from petroleum, they were first engineered from the blackness of coal tar. At the end of the plastic-making process, the blackness goes away and blankness enters, a pattern characteristic of the substance, since "nearly all plastics are, in the pure state, water-white, and in most cases highly transparent."[112] This processed evacuation of blackness into colorlessness bespeaks a new lifeless-deathless state. It mirrors Anthropocene *Man's* quest to imitate life, signaled in telling items like plastic flowers and all manner of "substitutes, analogues, imitations and duplicates, which because of the synthetic operations that bring them into being, seem to remain forever synthetic."[113]

Most plastics do not biodegrade, that is, break down and become transformed by bacteria and other organisms, which would enable their constitutive elements to re-enter the stream of life. They photodegrade, breaking down with exposure to light, but only into smaller, lingering, and consumable pieces.[114] Plastics thus achieve virtual immortality in bodies, oceans, and landfills, where they slowly leach sickening chemicals—"BPA, flame retardants, phthalates and poly-brominated diphenyl ethers (PBDEs)."[115] Ironically, plastic's refusal to rot, its single-use and throw-away convenience factor, and its colorlessness are what have made it so massively profitable.

That colorlessness is profitable because it allows what scholar of political aesthetics Esther Leslie calls a "synthetic rainbow,"[116] a spectrum of man-made colors (also originally derived from coal tar) to be poured into plastic's eerie blankness, making an otherwise unattractive substance desirable.[117] However appealing that artificial spectrum might first appear, Leslie discerns, it also disturbs for these are "overly intense colours, off-colours, odd colours, wrong colours . . . colours that shine too brightly, too intensely, colours that warp the prism, violate the rainbow and shatter the spectrum . . . colours from somewhere else . . . some educed in the imagination, some cooked up in the laboratory to outdo the natural world."[118] This violating rainbow, according to Leslie, epitomizes the goal of "industrial science: to reproduce the world synthetically."[119]

Two champions of the stuff, chemists W. G. Couzens and V. E. Yarsley, writing in 1941, prophesied a desirable plastic world to come, one that would expunge dirtiness, darkness, and decay and bring brightness and color—of the synthetic type:

> Let us try to imagine a dweller in the "Plastic Age" that is already upon us . . . "Plastic Man", will come into a world of colour and bright shining surfaces . . . a world free from moth and rust and full of colour . . . a world in which man, like a magician, makes what he wants for almost every need, out of what is beneath him and around him, coal, water, and air. . . . [W]e shall see growing up around us a new, brighter, cleaner and more beautiful world . . . the perfect expression of the new spirit of planned scientific control, the Plastics Age.[120]

At around the same time, *Fortune* magazine published an overview of the plastics industry that included two illustrations[121] that continue to stun with the virulence of their malice (masked as wonder). These reveal the gendered sexual violence informing the Plasticocene-Anthropocene as well as the use of color, or lack thereof, to represent the mutilation and murder of Mother Nature-Earth.

PLASTIC VENUS OF "SYNTHETICA"

The first of these illustrations is a multicolored relief map of a continent dubbed "Synthetica." Its nations, rivers, lakes, and surrounding lands are colored in different engineered hues, while individual places and waterways are named with different types and brands of plastics, for example, Nylon Island. A caption pays homage to those familiar Anthropocene ideals of limitlessness and *Man's* triumph over the "natural world":

> On the broad but synthetic continent of plastics, the countries march right out of the natural world . . . into the illimitable world of the molecule . . . Great chemical river systems, like the Acetylene, feed many countries . . . [T]he greatest plastic country of all—a heavy industrial region of coal-tar chemicals fed by Formaldehyde River is Phenolic.[122]

The map is horrific in many ways, including the name of that river. Formaldehyde, it turns out, is profoundly toxic to the "fish, shellfish and other creatures" whose homes are rivers.[123]

In his excellent study, *American Plastic*, historian Jeffrey Meikle introduces and analyzes the map of Synthetica, determining that it "firmly rooted plastic in the extractive materials culture of the past," recalling the exploitation of "civilization's colonial preserves." He apprehends the map as eerily resembling a medical anatomical model, "a graphic representation of earth's body opened up to reveal viscera of unnerving hues."[124] This description brings to mind Carolyn Merchant's reminder that the 17th-century Europe introduced a new language of "mastering," "managing," and "possessing" the Earth. This was illustrated with a "new image of nature as a female to be controlled and dissected through experiment."[125] In the map of Synthetica, I see not only that dissection of a corpse, but also the murder of the original living body. The map is a crime scene, specifically, a sex-crime scene.

This opened up body of land recalls the atrocities of the late 19th-century sex killer, Jack the Ripper. This patriarchal culture hero (now celebrated with his own museum in London) did not rape; he did not use his penis as a weapon. Rather, he used a weapon as his penis—a knife to slit the throats of five sex workers of the most stigmatized class in the poorest area of town. He then tore open their bodies, mutilating their breasts and genitals and even stealing the cunt-womb from one victim. Four of his murders were committed on the street, and the cut-up bodies were left on full display.[126]

"Ripperologist" Tom Cullen writes that the killer deliberately made his murders into "signposts" and asks, "What was the Ripper trying to say?"[127] Is it really that difficult to discern? The obvious messages are misogyny, the scapegoating of the most paradigmatically "impure" women, and still another way of fusing sex and violence.[128] This ritual murder signposted *Man's* extending victory over the stigmatized female body as well as the Earth body. This very recognition informs the classic 1963 film *Dr. Strangelove: How I Learned to Stop Worrying and Love the Bomb*, where a mad Air Force man jocularly named General Jack D. Ripper initiates world nuclear destruction.[129] No one has any trouble getting the joke. The map of Synthetica celebrates Plasticocene Man as serial sex killer/hero. It also names him master builder, able to make a new, cleansed, and artificially colorful (if poisoned) world literally over Earth's eviscerated body.

The second illustration accompanying the *Fortune* review of the plastics industry also depicts a conquered and possessed Mother Nature-Earth, using both artificial color and colorlessness. This illustration, "An American Dream of Venus,"[130] invokes the same goddess Henry Adams found in the Dynamo.

The title *Venus* has been applied to any number of Paleolithic Mother Nature-Earth figures, for example, the *Venus of Willendorf*. In ancient, pre-heteropatriarchal understandings, Venus is the Love or Sex deity not because she is the object of desire, but because, she "emits the desire that generates the energy of the universe."[131] Venus signifies "erotic love as a powerful and elemental aspect of nature."[132] As the Greek version of her myth goes, Venus (the Roman name for Aphrodite) was born from churning sea foam after Zeus castrated his father Uranus and threw the severed penis in the water.[133] Venus-Aphrodite was said to exude a glorious "fragrance," the very *funk* that stimulates sexual desire.[134] She also was and is *colorful* in Tomson Highway's sense of gender, as she is sometimes bearded. Venus is well understood as a deity who is all at once female, pansexual, pangender, femme, and transgender.[135]

The name Venus simultaneously, though, also conjures what Saidiya Hartman names as "ruin." She points to *Venus* as "a ubiquitous presence . . . in the archive of Atlantic Slavery,"[136] the name given by self-absolving abusers to countless women and girls they raped, murdered, and silenced, including on slave ships where they were then thrown into the sea. The name Venus also was affixed, Janell Hobson details, to Sarah Baartman, the Khoikhoi (South African) woman displayed in Paris and London as the Hottentot Venus.[137] Baartman was supposedly the very antithesis of Western norms, showcased in Botticelli's *The Birth of Venus* as a slim, sea-born blond beauty standing on a gigantic scallop shell.

With all this in mind, let's go back to *Fortune*'s "An American Dream of Venus." The Venus of this display is a transparent acrylic torso—headless, armless, legless, and without genitals. As in ancient Greek and Roman statues of women, her sex organ is missing, its place represented as a smooth, hairless, lipless, surface.[138] This acrylic Venus is an anti-erotic Sex Goddess—colorless herself, but surrounded by artificially colorful plastic objects. Meikle provides a lengthy description:

> Against a deep bluish-purple background of vinyl sheeting appeared the pellucid outlines of a stylized armless female torso rendered in acrylic. Within its reflective hollows . . . exotic goldfish swam and marine plants floated. Outside this otherwise undefined goddess, suspended about her in an immaterial medium, appeared the diffuse plastic accoutrements of her artificially colorful machine-age existence.[139]

Acrylic "Venus" is transparent: her darkness and her being denied, her mystery supposedly solved. Her heart, too, has been plucked. The space where that organ should be is occupied instead by a plastic playing card—a Queen, though not of Hearts but of Diamonds, signifying, perhaps, cashing in without compassion.

A *Fortune* caption rhapsodizes over the proliferation of plastics surrounding the dismembered goddess: "Dentures, doorknobs, gears, goggles, juke-boxes, crystal chairs, transparent shoes and ladies rise up from the plastic sea." Nearly a century later, "the plastic sea" (like the formaldehyde river) has been realized, horribly. Immeasurable quantities of plastic take up virtually permanent residence in oceans, on track to soon replace fish as the most common inhabitant of the planet's waters.

Swimming in the shallow waters of see-through Venus's nether region are plastic goldfish and some strands of seaweed. Meikle remarks, "With the exception of her own fishy depths, nothing suggested connection to living nature."[140] His observation is insightful and inspires me to dive deeper. *Fish* is slang for *cunt*, often uttered as a slur.[141] The reference is an olfactory one. In the "civilized" world carnality is deemed "savage" and smell is the most suspect of the five senses. A daily bath does not suffice. All natural bodily smells must be masked by frequently sickening, synthetic "fragrances" (also derived from fossil fuels).[142] The fishy smell of the vulva bespeaks humanity's origin in a uterine sea, our kinship with marine life, and the innate funk of the life-force. In Black gay drag culture, *fish* means an especially real drag queen.[143] The fake fish forever stuck in the depths of acrylic Venus openly lack the volatile essence of femme, funk, and female.

Meikle concludes that each *Fortune* illustration is "so bizarre in its own way, so rationally unwarranted, as to suggest an intrusion into consciousness from a site of unresolved psychological conflict."[144] He says no more on this, but to me that conflict smacks of cunt envy so profound it seems resolvable only in a final motherfucking of Nature-Earth. The artificially tinted land of Synthetica is *terra nullius*, remade as literally emptied of life and spirit. The "American Dream of Venus" transparently represents Plastic *Man's* ripper-like mutilation sex-murder of Nature-Earth.

The evidence is right there. Three strands of blood-red vinyl chains form a garrote around what remains of plastic Venus's neck. Screwdrivers, mechanical pencils, and pocketknives approach, indicating, Meikle says, "her male suitors."[145] But screwdrivers and such are "male" only in a violent phallic world where weapons pass as penis symbols because both are perceived, as Freud blithely put it, as having "the characteristic of penetrating into the body and injuring."[146] The sharp objects are not "suitors." They are rapists and rippers, rather like the men Tomson Highway describes who rammed a "screwdriver fifty-six times up" the vagina of a Native teenage girl, killing her.[147] Seized, dismembered, and plasticized Venus stands for rape-murder victims as well as *Man's* parallel violence against land.

The "American Dream of Venus" proclaims seizure of the Earth cunt-womb and the consequent ability to reproduce a new, substitute, synthetic world. The text accompanying the illustration extols the "limitless horizons, strange juxtapositions, and endless products of this *new world in process of becoming* (emphases mine)." Chemist and plastic enthusiast Edwin Slosson had proffered a similar prediction in 1919: "Man the Artifex" will "gradually . . . substitute for the natural world an artificial world, molded nearer to his heart's desire."[148] The "desire" here is a neo-necrophilic one—not a sexual desire for corpses, but one for objects, including living beings who have been turned into objects.

This neo-necrophilia is the "mystery" McLuhan senses behind the mechanical bride's cluster of "sex, technology and death." Musing further, McLuhan notes that "many people have become so mechanized" that they develop "a metaphysical hunger to experience everything sexually," to "pluck out the heart of the mystery for a super thrill."[149] Neo-necrophilia dis-spirits matter and then sexualizes the ensuing objectification. The targeted "heart of the mystery" is the primal cunt-womb that is Nature-Earth.

SOMEWHERE OVER THE SYNTHETIC RAINBOW

Around the world, the rainbow is understood as a bridge connecting the mundane world and the spiritual one.[150] Where, if anywhere, does the "synthetic rainbow" of the Plasticocene-Anthropocene lead? Some clues are given in the original Apple computer logo (1977–1998).[151] Colors akin to a striped, vinyl beach ball appear on the surface of a simulated apple with a bite taken out. These colors, though arranged in a rainbow-like pattern, are deliberately out of order. This synthetic spectrum neither glows nor pulses. And this apple has fallen so far, there no longer is a tree, let alone a green serpent or a Mother of all the Living. Once again, Genesis is the ur-text, updated for the Anthropocene. It's no longer necessary to demonize the serpent and Eve. Now, you can just delete them. The intelligence, knowledge, and nourishment—material and spiritual—that Earth gives is spurned. Instead, another sort of "intelligence," an artificial one, is sought—hence the bite from the vinyl-like apple.

Similar synthetic hues, and yet another homage to Genesis, appear on a *Wired* cover from 2015 that tells much the same story. The headline reads: "Welcome to the post-natural world." The scene celebrates what it calls the "Genesis Engine"—the technique of editing DNA to produce organisms by design—and predicts the outcome of its application: "No hunger. No pollution. No disease. And the end of life as we know it."[152] This engineered "post-natural world" is rendered with colors strange and out of place. The sky is still blueish, though with grey tones. More dramatically, the trees, growing sparsely on rolling hills, glow pinky-orangey, the ground pinky-purplish. There are no people or other animals. Like those tinting the map of Synthetica, these colors are unnerving. In reality, when colors are thus changed, as when a river turns bright orange,[153] it is a clear sign of poisoning.

If the colors of Earth do permanently change as a result of geoengineering, gene editing, runaway pollution, or some other interference, what would be the physical-emotional-spiritual consequences? In *Blue Mars*, Kim Stanley Robinson tells a sci-fi story about terraforming that planet for human settlement. At one point, the red Martian sky is made to look blue: "Terran sky blue, drenching everything for most of an hour, flooding their retinas and the nerve pathways in their brains, long starved no doubt for precisely that color, the home they had left forever."[154] Scholar Bronislaw Szerszynski cites this as part of his argument for the inclusion of environmental humanities in debates about geoengineering.[155] I don't disagree. Still, how much debate over the spiritual-physical-mental health benefits of nature's colors do we need? If we consult abuelita knowledge (La Loba Loca), probably not much.

Humanities scholars' contributions to debates on geoengineering would do well to teach histories like that of 1922 Rio de Janeiro. There a eugenicist, colonialist project murdered a mountain, displaced peoples, and filled the emptied space with a "white city," whose "perfection" was propagandized, again, with pictures of young, slim, and able-bodied White people, Europeanized landscapes, machines, and factories.

These humanities scholars also should include the history of specious promises of coming-soon techno-utopias, wrought by whatever newest thing is being sold to us by the world's richest corporations. Contemporary companies claim to provide, Franklin Foer argues, "an urgent social good, the precursor to global harmony, a necessary condition for undoing the alienation of humankind." But behind this utopic PR, he perceives, is a desire to "implant their artificial intelligence within our bodies . . . to complete the long merger between man and machine—to redirect the trajectory of human evolution."[156] Foer warns that humans are not so much merging with machines as, instead, being assimilated into these corporations.

Recognizing this does not mean that to be environmentally, politically, and spiritually sound humans must never work with machines, pursue new technologies, or use plastics. The concern is to do it wisely. Most obviously, plastics are needed for medical uses that save and enable people's lives. The use of technology and plastics to assist those with disabilities, eco-ability scholars perceive, actually *is* ecologically sound because it allows disabled people "to be self-reliant and reinforces that dis-ability is a valued quality, which should be respected and praised. This assistance stresses the ecological importance of interdependency on which the life system is based."[157]

One scholar of science and technology studies, Jody A. Roberts, named himself a "plastiphobe" and practiced avoiding it. But then one of his children was born with cerebral palsy, and immediately required tubes and other plastic materials to sustain her life. Roberts, while remaining aware of plastic's harms, affirms the need to resist "simple dichotomies." Plastics "are not simply life saving or a threat: they are both."[158] Green scientists are working on the development of a more benign plastic-like substance that will save some lives without endangering others. Wise use, Roberts insists, requires the establishment of limits on "how and when and why we use"[159] any plastics—now and thirty thousand years from now.

Alice Walker avers that "even tiny insects in the South American jungle know how to make plastic . . . they have simply chosen not to cover the Earth with it."[160] The truth underlying her flippancy is that all Earth beings create, think, make, and shape the world. It's human exceptionalism that trains us to dismiss manifest evidence (like cyanobacteria inventiveness, birdsong, and termite towers) of other-than-human beings' ingenuity, creativity, and culture, and to classify these as "instinct" and not intelligence. That same sort of dismissal marks colonialist mindsets when confronted with Indigenous peoples' scientific achievements, creativity, and cultures. The pertinent questions are whether or not you invent and produce wisely. Do you "look to the mountain" (if it still stands) and consider how the new development will continue to affect others thirty-thousand years into the future? Do you take into account the ways that your reshaped environment and the machines you rely on will inevitably shape you back?

The cover of *The Economist* from February 5, 2015, announces the "Planet of the phones" and shows the Earth from space, made up entirely of cell phones. Each screen displays something colorfully earthlike—a bluish sky, a sandy piece of land. But the joke is on phone-dependent viewers, because no one could survive on this planet. Even now, cell phone technology is shaping us in ways we might not expect or want. Sociologist Sherry Turkle, for one, cautions that phone immersion diminishes human empathy, introspection, creativity, community, and intimacy, all of which are dependent upon face-to-face relationships.[161]

Our use of technology without careful consideration of the consequences, in the eyes of eco-philosophers, is the epitome of "insidiousness." This is "the idea that communities that adopt technological hardware, such as television, or rely on technological supply chains, such as global agrifood, cannot stop the erosion of their previous, more intimate relation with the environment."[162] Insidiousness "creates the illusion that humans are not dependent on and cannot be affected by Earth systems such as the climate system. It alienates people from the combined impacts on the environment of their individual actions."[163] Mid-20th-century eco-philosopher Aldo Leopold also remarked upon this phenomenon, finding acute "spiritual danger" in the belief that "breakfast comes from the grocery."[164]

Nowadays, insidiousness includes the ways people in affluent nations use plastic routinely, thoughtlessly, and then just throw it "away." The word *plastic* signifies something that can be shaped into desired forms, embodying "the very idea," Roland Barthes writes, "of infinite transformation" and "the euphoria of a prestigious free-wheeling through Nature."[165] Do any of us who are responsible for plastic's unlimited production and use consider how that "free-wheeling through Nature" smacks of rapism, how plastic, broken down into micro-particles or releasing harmful chemicals, silently invades bodies around the world, human and other than human, without their consent? The Anthropocene's essence is captured by Pamila Gupta and Gabrielle Hecht in a visualization worthy of a horror film: "There's now enough concrete on the planet to produce a 2mm thick, full-scale replica of Earth, and enough plastic to completely wrap that replica in cling film."[166] Does such an accomplishment really sound like *Man* has outdone nature and become a "God Species"? Or, does it suggest quite the inverse?

Those geoengineered, pinkish-leafed trees on the "Genesis Engine" cover of *Wired* illustrate what the headline calls the "end of life as we know it." Those same color codes sometimes can signify freedom, queerness, and the end of heteropatriarchy, which many of us would like to see end. But that is not what is going on here. Like the pure white space of "Globopolis" and the simulated colors on the screens of "planet of the phones," this dissonantly colorful geoengineered neo-genesis looks habitable only by machines, extraterrestrials, or fabulously adaptable bacteria. There might be a temporary enclave somewhere in this neo-necrophilic playground, where rich human immortals plan a party to toast their neo-Genesis. My guess, though, is they find that the party is only a success when thrown by the Black Madonna and her colorful entourage. When this realization hits them, they may try, unsuccessfully, to suck sustenance from the always-dry breast of plastic Venus, to crawl back into the empty space of her eviscerated womb, and to beseech this heartless sex object to bestow, once again, the gift of death.

Those uncanny sister simulacra—the heart-plucked mechanical bride, the funk-less fembot, and the cunt-less Venus—all signify something beyond the actually comforting realities of life, darkness, blood, smell, sex, decline, death, and rebirth. This trinity of transparent Sterility Goddesses bespeaks collapse with no guarantee of renewal. They come to the fore when, faced with disrespect, Mother Nature-Earth takes her leave.

I saw something like this once in a dream, one that opened with me receiving a post card in the mail.[167] An unsigned message suggested that I visit a park to behold what the card pictured—a great deciduous tree with many branches and a full canopy of leaves. Eerily, someone had carved an exact replica of the tree into its uppermost trunk. I went to the park and finally spotted the tree at a short distance, where I took a seat on a bench and observed. At the base, a pair of male-female lovers circled the tree but could not reach out and touch the living surface. Smooth, grey concrete blocks had started to surround the bark of the trunk from the ground up. The carved replica on the upper trunk still was visible, but it seemed only a matter of time before the blocks covered even this. The dream seems to suggest that as *Man* continues his quest to "substitute for the natural world an artificial world, molded nearer to his heart's desire" (Slosson), the colorless consequence is only ever-more-profound estrangement from the Tree of Life, from paradise, from the "Mutha'" Funkin' Earth.

INTERLUDE BETWEEN A CURSE AND A SWEAR

This section is inspired by *motherfucker* in its capacity as a "swear word." A *swear*, like a curse, invokes a deity but now with the intent of encouraging that deity "to empower your words." The deity called on here, once again, is Mother Nature-Earth.

These chapters answer concerns raised in the earlier *Curse* section, proceeding in a reverse or ritually unwinding way. Chapter 7 responds to the color loss tracked in Chapter 6 as it considers the spiritual meanings of the color green. Chapter 8 breaks the silences imposed by Anthropocenic motherfucker culture, flipping its script of omnipotence to land on cunctipotence and attending to the Word continuously birthed by the "Mutha'". Having denied worship to the tower-of-power god(s), Chapter 9 gathers some of the visions and stories now calling, and calling on, the "Mutha'".

THE SWEAR

"from Old Norse svar answer, svara to answer." (*OED*)

"FEED THE GREEN"

IN AWIAKTA'S ADMONITORY POEM "WHEN Earth Becomes an 'It,'" Mother Nature-Earth, having been met with disrespect, turns her face and takes her leave: "She is taking all green/into her heart/and will not turn back/until we call her/by her name."[1] This chapter invokes that mothering power as well as the *greenness*. The title is in quotes because the phrase *feed the green* is not mine. That imperative came through in a dream in October 1999; it woke me up and I wrote it down.

The dream opens in the Boston metro area, where I lived in the 1970s. I am rushing to get to an aboveground "T" station, which is, funnily enough, along the Green Line, so named in the city transit's color-coded system. When I arrive, I spot a door that usually isn't there. Along with a few others, I enter and find a staircase, which we follow into a world beneath this one. There we find a train station and people, but they are not human. These folks are not too pleased to see us, having found humans to be unmannered, though no one makes any move against us. I figure it best to make myself inconspicuous, so I get on a train, and stay quiet and small. After some time, the train stops and several of us disembark into a sun-bright and verdant countryside. We walk a short way and come upon a spring welling up from the ground, a flow of sparkling greenness. I dip my arm into the stream, feel a surge of energy, and, withdrawing my arm, find it now to be all green and aglitter. The dreamscape then shifts abruptly, and I am outside in New Mexico, where I also once lived, standing alongside a friend, a Pueblo woman. I ask her, "What does it mean?"

Call Your "Mutha'". Jane Caputi, Oxford University Press (2020). © Oxford University Press.
DOI: 10.1093/oso/9780190902704.001.0001

She points to an expanse where she is sowing seeds for three local grasses that birds love to eat, looks at me, and replies, "Feed the green."

The feelings and perceptions that are communicated through a dream often impel the dreamer to act or change in some way. This one awakened in me a strong desire to find out all I could about *greenness*. Obviously, green is the color of the environmental movement and of plant life. But what further spiritual meanings does green convey? Why did this green spring sparkle with its own internal fire? Why should we and how might we feed this *green* that so manifestly feeds us?

Green (along with blue, often indistinguishably) is the color of water[2] as well as vegetation. Photosynthesis was first achieved by cyanobacteria (who, along with billions of other microorganisms, actually do live in a world underneath this one).[3] Plant photosynthesis is powered by green chlorophyll interacting with the radiant orange-yellow-redness of the sun. The process, as science journalist Oliver Morton puts it, enables plants to "eat the sun."[4] They then become food for others, who are, in turn, eaten and returned to the elements. Green plants imbibe fire and then give both fare and air. To be on good terms with life, we are obliged to return the favor, in essence, feeding the green.

Morton writes that green stands out among all colors because "the greenness of life is so important and all-pervading that evolution has turned our eyes to discriminate among its various hues more precisely than among those of any other color, and so shaped our brains that we take solace in it. The green, we know without thinking, is good."[5] This is a complex goodness. Green, again, is the color of life, but also the color of slime, sickness, snot, vomit, decay, and death—all of which seem grim. Still, a "good death" (not the premature and torturous deaths determined by oppression) is good in that it enables transformation and renewal. In the Anthropocene, though, new and not so good (at least for us) greens have arrived on the scene—weird greens, disturbing greens, toxic greens, excessive greens, out-of-place greens, mutagenic greens.

They convey materially (e.g., toxic blue-green algae blooms)[6] and symbolically (e.g., rampaging green monsters like Godzilla) the incommensurable changes going on in the Anthropocene.

In Pueblo philosophy, human beings are tied to land. Poet Simon Ortiz (Acoma) explains that because the land is the very source of life, humans are obliged to "give life back to it."[7] N. Scott Momaday recounts a related Kiowa practice. His father was visited regularly by an old man, who "would get up in the first light, paint his face, go outside, face the east, and bring the sun out of the horizon. Then he . . . would pray aloud to the rising sun. He did that because it was appropriate . . . the sun was the origin of his strength. He understood the sun, within a more formal religious context, similar to the way someone else understands the presence of a deity. And in the face of that recognition . . . [t]hrough the medium of prayer, he returned some of his strength to the sun. He did this everyday."[8]

Returning power is necessary because the creation of life is not something that happened once and then stopped. Creation is ongoing. Kenyan Green Belt Movement founder and Nobel laureate Wangari Maathai affirms our obligation to "replenish the Earth."[9] This is done through every aspect of being and living, personal and collective, and includes tangible actions (like the Green Belt planting of trees) along with energetic rituals of renewal, including the outpouring of prayer, attention, love, funk, strength, respect, and gratitude. These rites of renewal, if you are from a culture that doesn't already have them, cannot be appropriated from other traditions. They need to be sensed, heard, and brought into being.

Scholars Raj Patel and Jason Moore, in their critique of the Capitalocene, cite Winona LaDuke's reference to "*mino bimaati-siiwin,*" meaning "continuous rebirth."[10] They lament that "there isn't a word in English for the process of making life" and contrast this with the Southern African Bantu word *ubuntu*, meaning "human fulfillment through togetherness," and the Shona word *ukama*, "a relatedness to the entire cosmos."[11] Actually, there is a phrase in African American English, "everythang is everythang," which "suggests the connectedness or oneness of people, places, events."[12] *Feeding the green*, for me, is still another way of naming that underlying oneness and reciprocity required for the process of making life. Such renewal can and must take many forms, including making art, building community; inventing green technologies; collective organizing for justice; restoring land; planting seeds; tending to, caring for, and paying attention to others, including other than humans; and sharing meaningful dreams and visions. According to witnesses, Martin Luther King Jr. did just that to great effect. In 1963, at the March on Washington, he began by reading his prepared notes, but was exhorted by singer and "Queen of Gospel" Mahalia Jackson to instead "tell them about the dream, Martin," at which point he set aside the text and did just that, resulting in his famous "I Have a Dream" speech.[13]

It is not only remarkable people like King who dream meaningfully. Like any gift, a green dream comes with obligations. Leanne Betasamosake Simpson, in an interview with Naomi Klein, tells her that in "Anishinaabeg philosophy, if you have a dream, if you have a vision, you share that with your community, and then you have a responsibility for bringing that dream forth, or that vision forth into a reality. That's the process of regeneration. That's the process of bringing forth more life—getting the seed and planting and nurturing it. It can be a physical seed, it can be a child, or it can be an idea. But if you're not continually engaged in that process then it doesn't happen."[14] *Feed the green* is a dream that I want to share.

BEING GREEN

In folklore, popular culture, the arts, literature, sacred texts, and eco-activism, greenness is everywhere.[15] This virid panorama appears in distinct traditions from different places, and I trace a few examples here. I do not mean to ignore their real differences, nor to claim identical meanings among all, but I believe that however diverse and local, they also might share a symbolic rootedness in "the greenness of life" (Morton).

In art, Mother Nature-Earth is nearly always seen in—and as—green, often with a dash of red. Green and red are complementary colors as there is no trace of one in the other; they enact the "reconciled opposites"[16] basic to the recipe for making life. A 1949 painting by Frida Kahlo, *The Love Embrace of the Universe, the Earth (Mexico), Diego, Me, and Señor Xólotl* (Figure 6.1) shows an ambiguous green and brown Universe as the enveloping background, in front of whom green-brown Mother Earth appears. Kahlo, in a red dress, sits on her lap, and holds in her own lap a naked, baby-like Diego Rivera. The dog, Señor Xólotl (suggesting the psychopomp, able to cross lands of living and dead) snoozes on the side. Green Mother Nature-Earth also regularly shows up in activist graphics, as in Lisa Cowan's original "Mother Earth is a lesbian" button (© White Mare, Inc.) that shows no human figure but deeply verdant trees and grass.[17] Mother Nature is green and with a red and fiery aspect in Disney's environmentally oriented though, unfortunately, culturally colonialist *Moana*.[18]

In explicitly sacred imagery, green beings evoking an Earth-based deity abound. Xochiquetzal, the Aztec goddess of sexual desire, grows green horns. The titles of the Sumerian Inanna include "She of the springing verdure" (in other words, a green fountain) and "The Green One."[19] That last is also a title of Demeter, the Greek Grain Mother.[20] Demeter's daughter-self, Persephone, is green and spends time in the underworld to be "regenerated by chthonian red"[21] of the pomegranate. The story of her abduction and rape by Hades is a later one, reflecting the violent transition to patriarchy.

In ancient Egypt, *Wadjet*, "the green one," was the name of the protective serpent Goddess of Lower Egypt, while "to do 'green things' was a euphemism for positive, life-producing behavior."[22] The deities Hathor, Isis, and Osiris all are associated with green, and Isis is known as the "'Creator of green things,' 'Queen of

FIGURE 6.1 Frida Kahlo, *The Love Embrace of the Universe, the Earth (Mexico), Diego, Me, and Señor Xólotl.* © 2019 Banco de México Diego Rivera Frida Kahlo Museums Trust, Mexico, D.F. / Artists Rights Society (ARS), New York.

Earth,' and the 'Green Goddess, whose green color is like unto the greenness of the earth.'"[23] Egyptian founding feminist Nawal el Saadawi titles her autobiography *A Daughter of Isis*, and delights in her characteristic color: "My eyes are drawn to green, my lungs breathe it in. Green is converted in my lungs to oxygen."[24]

Islamic traditions speak of "Khidr, Khisr or Al Kadir, the Green Man," who signifies "the mean" or middle in relationships. In one story about him, Khisr travels across the desert carrying a dead fish and finds a wondrous spring. He then "dipped the fish in the water and it immediately came to life again. Khisr realized that he had found the fountain of life. He dived in and became immortal and his cloak turned green."[25]

A Buddhist "universal mother" Tārā, scholar Miranda Shaw writes, "is envisioned as green in her original and most enduringly popular form."[26] One devotee praises her: "Your body is emerald green, blazing with splendor, invincible./If one simply remembers your name, you come forth to save from/any terror."[27] Shaw points to green Tārā's culturally specific meanings in Buddhism, but also connects her "green hue" to "an association with nature, trees, and vegetation, raising the possibility that she embodies magical and religious properties connected with plant life."[28] Other green and greening deities include Venus (with grass springing up around her every step), Hindu's Vishnu, the European pagan Green Goddess, and the semi-divine European folk figure of the Green Man, whose face can be discerned amid foliage. Its painful to look at the porno-tropical paintings of 19th-century Frenchman Paul Gauguin (who imposed sex on Polynesian teenage girls), but he did paint the dead Christ as green, linking his story of death and resurrection to that of other green deities reborn with the springtime.[29]

In European folk culture, green garb and a red cap and shoes are associated with fairies, elves, and leprechauns, who can be sometimes helpful, sometimes tricky.[30] Fairies from the Blackdown Hills in Somerset, England, dance and sing "Green, green, green, All a-green, all a-green/ A-dancing round the tree."[31] Another song tells, "In yonder greenwood a green lady doth dwell, Her hair it is green and all green is her gown, And she calleth to all, draw near, come here."[32] Writer Malidoma Patrice Somé reports an encounter with a green woman dressed in black in a yila tree during an adolescent initiation ritual in Burkina Faso. This woman's skin color "was green, light green." Still, he apprehends that her "greenness . . . had nothing to do with the color of her skin. She was green from the inside out, as if her body were filled with green fluid. I do not know how I knew this, but the green was the expression of immeasurable love."[33] Africana spiritual traditions regularly feature "trees as spiritual mothers."[34]

Trees come "eventually to symbolize almost every major mythical figure,"[35] especially Mother Nature-Earth as the Tree of Life. Sophia, the personification of Wisdom in the Gnostic texts, is both a tree and "the Womb that gives shape to the All."[36] In Ecclesiastes, Sophia announces, "I have grown tall as a cedar . . . a Cyprus . . . a palm . . . Approach me, you who desire me, and take your fill of my fruits."[37] This generosity, though, must be reciprocated or the welcome can be withdrawn.

The Pomo language has a word for this obligation to the green. As one spokesperson told an anthropologist in the early 20th century, "Plants are thought to be alive, the juice is their blood, and they grow. The same is true of trees. All things die, therefore all things have life. Because all things have life, gifts have to be given to all things. This is called *gaXol cayoi* (outwards gift)."[38] The classic American children's story *The Giving Tree* teaches a very different and damaging lesson. This picture book has an eye-catching green cover, featuring a little White boy in a green T-shirt and red overalls, accepting a red apple from a green tree. As the boy matures he becomes unloving, distant, and self-serving, taking terrible advantage of the clearly feminine-identified tree.[39] The tree gives and gives, never expecting anything in return. By story's end, the boy-now-man has dismembered her, and all that remains is a saddened stump. At the end of his life, the boy returns and sits on the stump, making the tree again "happy." This terrible tale transmits outright motherfucker mythos, updated in the 2017 film *Mother!*[40]

Representatives of Mother Nature-Earth who refuse self-sacrifice are villainized in pop culture, notably Batman's red-haired and green-skinned nemesis Poison Ivy.[41] Consider, too, the infamously green Wicked Witch of the West in *The Wizard of Oz*, rightfully desiring the return of the red shoes. In contrast to the charlatan Wizard, the Witch actually does have magic powers and flying monkey friends. Eventually, the envious Wizard uses Dorothy to get rid of his rival. But the Witch's glorious greenness is revived as she becomes an animal-rights and anti-fascist activist in Gregory Maguire's novel *Wicked* and the wildly popular Broadway musical based upon it.[42]

Another fan favorite is Maleficent in Disney's *Sleeping Beauty* (1959).[43] This green witch abides on a clitoral hill, carries a phallic staff with a green glow at the tip, and wears a horned crown and cape of black and purple. She casts spiraling airborne spells, has a crow as a companion, envelops herself in green fire, and shape-shifts into a green-fire-breathing dragon. She is slain by a forgettable prince-hero, which apparently ends the threat to the kingdom. The implicit promise though, is that, someday, Queendom will come.

The infamous monster Dracula is green in many depictions, including a poster for a theater production of the novel from 1928. This Dracula is brilliantly green, with enormous, plant-like bat wings for ears as well as a fanged red mouth, red-rimmed eyes and lips, and a wrinkly, bony forehead that suggests the shape of a great tree.[44] [Figure 6.2] Green Dracula hails from the same stigmatized kin group as the European witch, characteristically depicted as a hybrid of animal, plant, and human, who often has animal companions. A 16th-century German woodcut showed the witch as a naked, aged woman who is bat-winged, horned, and wrinkly, like tree bark, with a serpent (presumably a green one) winding round her left leg (Figure 6.3).[45]

The green serpent in Genesis is another version of Mother Nature-Earth made bad. Artist, color enthusiast, and

FIGURE 6.2 Poster of the play *Dracula* (1928). Heritage Auctions, HA.com.

gay rights activist Derek Jarman identifies the snake as the green and "ancient Venus, old enough to be God's granny," while he also bemoans the "colourless world" imposed by the "unpleasant new God."[46] Surrealist Leonora Carrington's

FIGURE 6.3 Georg Pencz, illustration for Hans Sachs, *Das Feindtselig Laster, dey heymlich Neyd* (Nuremberg, 1534).

women's liberation poster from 1972, *Mujeres Conciencia* (*Women of Conscience*; Figure 6.4) is based on the garden of Eden story, but now reclaims the earthy paradise and the gift of knowledge bestowed by nature. Eve and Adam are two

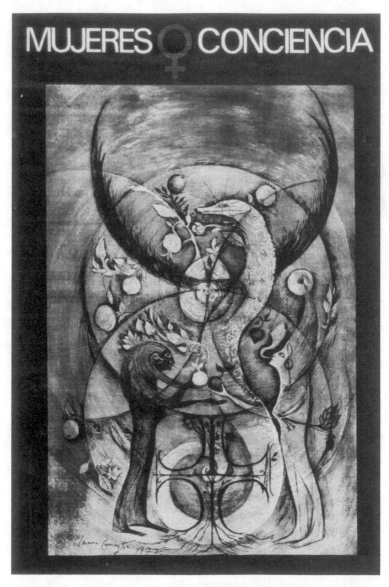

FIGURE 6.4 Leonora Carrington, *Mujeres Conciencia* (1972). Gouache on cardboard, 29' ½" x 19' 3/10", 74.9 x 49 cm. CREDIT: © 2019 Estate of Leonora Carrington / Artists Rights Society (ARS), New York.

green Egyptian female figures with the creative power of the green serpent rising between them, winding round the cosmic Tree and offering that gift of green consciousness in the form of the apple.[47]

Alison Saar celebrates the green serpentine power conjoined with the Tree of Life in her mixed media artwork *Mamba Mambo*.[48] A sculpted brown-skinned Mambo (a Vodou priestess) stands on a green pedestal placed atop a high green stool. Curling around her tree-like body is an enormous green mamba, a venomous southern African cobra. The Mambo's bright red lipstick and matching high heels, art historian Lisa Farrington writes, constitute "the artist's code for the female element."[49] A flag painted with two identical palm trees hangs behind the Mambo sculpture. Like the Gnostic Sophia, the Mambo is as tall as the trees, with sun-like rays emanating from her body. Through this work, Saar pays homage to Vodou's non-binary, unifying, and often green-toned deities "Dambalah and Ayida Wèdo, who are represented in Vodou art as double snakes coiled around a palm tree."[50] (See Figure 6.5.)

All this greenness—from the pop to the profound—affirms Morton's assertion that humans "don't just enjoy seeing the green," but need the green and must heed the green, for green is the color that "shapes the possibilities of our lives."[51]

GREEN PLACES

Reading about Khisr's dip in the green fountain as well as Inanna's title, "She of the Springing Verdue," I was reminded of the virid spring in my dream as well as a passage from the *Tao Teh Ching*:

> The subtle essence of the universe is active.
> It is like an unfailing fountain of life
> which flows forever in a vast and profound valley.
> It is called the Primal Female and the Subtle Origin
> The Gate of the Subtle Origin.[52]

FIGURE 6.5 Alison Saar, *Mamba Mambo*. © Alison Saar. Courtesy of L.A. Louver, Venice, California. Mixed media, 5'4" x 1'5" x 2'10" x 1'5".

An alternative dimension with a cosmic gate and still another greening fountain shows up in a Marvel-comic film *The Fantastic Four* (2015),[53] in which youthful scientists travel through a "Quantum Gate" to a tumultuous other world where a glowing green liquid flows up from the ground. One adventurer dips his arm into the radiant green fountain and the ground erupts, seemingly killing him, though he comes back as the green-cloaked villain, Dr. Doom. Perhaps the writers knew of Khisr's experience with the inexhaustible green fountain of life and similarly clad their villain. Or maybe they had read the *Tao Teh Ching* and knew of the green valley and the unfailing fountain. Or perhaps, the green, transformative spring just came to them, as it did to me, because it is a universal symbol. Perhaps, even, this green spring or fountain is something of a real presence. Environmental philosopher Aldo Leopold, in his classic essay, "The Land Ethic," (1949) defined *land* as "not merely soil," but "a fountain of energy flowing through a circuit of soils, plants and animals."[54] The fabled fountain of "immortality" is real in this sense of the inexhaustible, ongoing, and renewing life stream.

In the explicitly imaginary world of DC comics, "The Green" *is* a real place, one formed by plant energy:

> A transcendent state of harmony with all of nature in which the "knower" is united with the "known." The energy of all living vegetation forms the Green, which extends as far as plant life reaches. It is a restorative, healing place where there is compassion and love for all. One who enters the Green feels he is slipping into a cool and comfortable place where all cares dissolve into the safe and nurturing bosom of the Mother of All Life.[55]

This passage may seem but a flowery, even infantile, fantasy. Still, this "Green" is akin to Martin Buber's philosophy of the beneficence of "I-Thou" relationships, ones steeped in recognition, respect, and reciprocity.[56] For reasons ranging from trauma to power-madness, oppressors enforce "I-It" relations, treating another being as an object for instrumental use.[57] They have lost access to "the Green." They do not feel *ukama*—their relatedness to the entire cosmos,[58] allowing ecocide among other devastations.

Toni Morrison communicates the spiritual-energetic power of the green place in *Beloved*. "The Clearing" is a spot in the woods where Baby Suggs led profound rituals, a place where "boxwood bushes, planted in a ring, had started stretching toward each other . . . to form a round, empty room,"[59] "a green blessed place . . . misty with plant steam and the decay of berries"[60] (red ones probably). Years later, Baby Suggs's surviving granddaughter, Denver, finds that regenerative place. She bathed in its "emerald light . . . Veiled and protected by the live green walls, she felt ripe and clear, and salvation was as easy as a wish."[61]

In *Rooted in the Earth: Reclaiming the African American Environmental Heritage*, Dianne D. Glave cites Morrison and affirms the greening influence of Africana spirituality: "For African farmers, the cycles of planting and harvesting, alpha and omega, and life and death were fundamental to nature and agriculture. Farmers consecrated and benefited from the soil, which was, in turn, a source of spirituality, nourishment, and life to humans." In these ways, "Africans blended the sublime with the commonplace."[62] This combined caring for and consecration of land *is* feeding the green. Glave recommends that African Americans build on this legacy, including by taking "worship outdoors onto the lawns and in the midst of trees on church property" or to "nearby parks or trips to wilderness places . . . God waits inside the church. S/He is also waiting by a waterfall or under a tree for communion and fellowship with those of us here on earth."[63]

Taking worship outdoors to green places, with care to make it accessible to all, is a mode that can be adopted by any tradition. As a young girl, raised as a Catholic, I had little interest in going to church but would often take a discarded shoebox and make altars to the Madonna, always outside. Gifted individuals like ecofeminist theorist and activist Starhawk revive European pagan traditions and conduct rituals outdoors, including ones marking the solstices and worshiping dirt.[64] Referencing Okangan educator

Jeanette Armstrong's instructions, Starhawk recommends a personal practice of finding at least one place to know intimately, and then work to develop and nourish that bond.[65]

Carol Lee Sanchez (Laguna Pueblo) mourns that so few non-Indigenous peoples bond with the land: "Dominant Euro-Americans waste the resources and destroy the environment in the Americas because they are *not* spiritually connected to this land base, because they have no ancient mythos or legendary origins rooted in this land" and also, she says, because they believe that their God authorized them to have dominion over a "profane" land. Sanchez wants non-Indigenous peoples to "acknowledge and become thoroughly familiar with the indigenous spiritual frameworks of *this* hemisphere and establish a connection to *this* land base where we were born."[66]

Robin Wall Kimmerer suggests that "for all of us, becoming indigenous to a place means living as if your children's future mattered, to take care of the land as if our lives, both material and spiritual, depended on it."[67] To do this with integrity, Leanne Betasamosake Simpson further advises, requires establishing real relationships with Indigenous people, else this would continue the colonialist stratagem of extractivism: "Extracting is stealing. It is taking without consent, without thought, care or even knowledge of the impacts on the other living things in that environment." An extractivist worldview means defining everything and everyone Indigenous as a "resource." There is, Simpson continues, an "alternative to extractivism" and that "is deep reciprocity. It's respect, it's relationship, it's responsibility, and it's local," and it includes people living within their own "fifty-mile radius."[68] This may be hard to fathom for many of us, myself included, as the lifeways of many Americans are very far from this green way of being. It's a necessary challenge to reckon with, including by taking personal responsibility and by participating in community efforts to get there.

FEEDING GREEN FIRE

The green fountain in my dream glittered with an internal fire.[69] Such flaming greenness has long been popularly associated with magic as well as European alchemical tradition, whose practitioners sought the power to copy nature by gaining access to Earth's "'secret fire, a living and radiant spirit'" taking form as a "green, translucent crystal."[70]

The 12th-century German theologian, abbess, composer, and healer Hildegard of Bingen made up a Latinate word—*viriditas*, literally "greenness," to name "all the life-giving qualities of God's spirit in matter, both human and non-human."[71] Hildegard also exalted the Madonna as a fiery greenness, bringing the whitened Virgin back down to Earth and recognizing her as both green and blazing:

> What greater praise can I give you than to call you green? Green, rooted in light, shining like the sun that pours riches on the wheeling earth; incomprehensible green, divinely mysterious green, comforting arms of divine green protecting us in their powerful circle. And yet, lady, you are more than even the noblest green, for you glow red as breaking dawn, you shine white as the incandescent sun. Splendid virgin, none of our physical senses can explain or comprehend you.[72]

Another translation of this passage has Hildegard praising the green Madonna as "burn[ing] like a flame of the sun"[73]—much as Green Tārā's devotees find her "blazing with splendour." That same congruity appears in Tanya Torres's painting *Atabey Gives Birth to the Coquí* (frog), showing the Taino Mother of All as reddish-brown skinned, with glowing green hair. Green leaves in the shape of flames rise before her (Figure 6.6). Atabey, Torres explains, is giving birth via "a fire that does not burn."[74]

FIGURE 6.6 *Atabey da la Luz al Coquí* by Tanya Torres ©2012. Oil on Canvas, 20" x 16".

The paintings of Edouard Duval-Carrié carry the cosmo-vision of Haitian Vodou, which includes knowing all elemental beings as alive and enspirited. In *Primitif Futur*, flickering orange flames enclose verdant foliage. In their midst emerges a being with bright eyes and lips, intrinsic to the surroundings (Figure 6.7).[75] The work invites an epiphany, a sudden and intense awareness of the green commonplace as brimming with meaning and presence.

FIGURE 6.7 Edouard Duval-Carrié, *Primitif Futur*. Mixed Media in Artist Frame, 33" x 41" (2002). Photo: Carl Juste. Private Collection.

I asked green religious studies scholar Bron Taylor about all this recurrence of fiery greenness, and he pointed me to Aldo Leopold and his signature vision of "green fire."[76] Leopold, originally from Iowa, had directed the US forest service in New Mexico and Arizona. His evolving ideas are closely aligned with Indigenous philosophies and practices, though he does not cite this influence.[77] In "Thinking like a Mountain" (an idea that is reminiscent of the Tewa practice of "looking to the mountain" described by Cajete) Leopold recalls a life-changing moment he had experienced decades earlier in Arizona at midday. On a mountain with a group of other men, he encountered and reflexively opened fire on a mother wolf and pups, killing her and most of her children:

> We reached the old wolf in time to watch a fierce green fire dying in her eyes. I realized then, and have known ever since, that there was something new to me in those eyes—something known only to her and to the mountain. I was young then, and full of trigger-itch; I thought that because fewer wolves meant more deer, that no wolves would mean hunters' paradise. But after seeing the green fire die, I sensed that neither the wolf nor the mountain agreed with such a view.[78]

US "game management" policy at the time allowed routine wolf-murder, with an ultimate goal of species extirpation.[79] Leopold's encounter with the mountain's knowledge and the wolf's extinguished green fire made him realize the horror of this ongoing destruction of biological integrity. With no wolves, deer multiply and eat so much vegetation that the mountain erodes. Then the deer herd dies off, "dead of its own too-much," while "dustbowls" prevail and rivers wash "the future into the sea."[80]

Green fire went on to become a catchphrase for ecological concerns, including in the title of a book about photosynthesis by two Spanish scientists.[81] Finding it odd that there was no

FIGURE 2.1 Amir Khadar, *Trans Day of Resilience Poster* (2017). Digital Illustration & Collage.

FIGURE 6.1 Frida Kahlo, *The Love Embrace of the Universe, the Earth (Mexico), Diego, Me, and Señor Xólotl.* © 2019 Banco de México Diego Rivera Frida Kahlo Museums Trust, Mexico, D.F. / Artists Rights Society (ARS), New York.

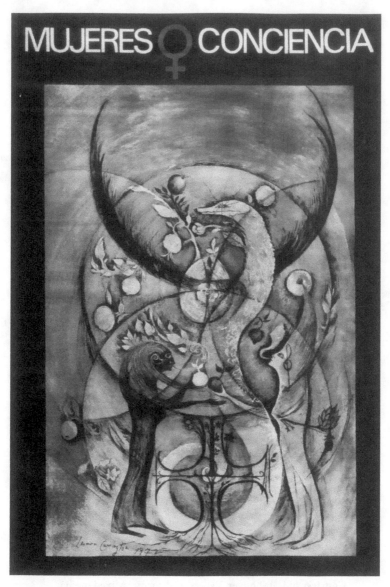

FIGURE 6.4 Leonora Carrington, *Mujeres Conciencia* (1972). Gouache on cardboard, 29' ½" x 19' 3/10", 74.9 x 49 cm. CREDIT: © 2019 Estate of Leonora Carrington / Artists Rights Society (ARS), New York.

FIGURE 6.5 Alison Saar, *Mamba Mambo*. © Alison Saar.
Courtesy of L.A. Louver, Venice, California. Mixed media,
5'4" x 1'5" x 2'10" x 1'5".

FIGURE 6.6 *Atabey da la Luz al Coquí* by Tanya Torres ©2012. Oil on Canvas, 20" x 16".

FIGURE 6.7 Edouard Duval-Carrié, *Primitif Futur*. Mixed Media in Artist Frame, 33" x 41" (2002). Photo: Carl Juste. Private Collection.

FIGURE 8.2 Detail from John Thomas Biggers, *Web of Life*. © 2019 John T. Biggers Estate / Licensed by VAGA at Artists Rights Society (ARS), NY, Estate Represented by Michael Rosenfeld Gallery.

FIGURE 8.4 Earth Mother Goddess, detail from Kevin Sampson, *Fruit of the Poisonous Tree*.

FIGURE 8.5 Wangechi Mutu. *The End of Eating Everything*, 2013. Animated color video with sound. 8 minutes, 10-second loop. Copyright Wangechi Mutu. Courtesy of the artist, Gladstone Gallery, New York and Brussels, and Victoria Miro.

mention of Leopold in this book, I contacted the editor, John Oakes. He told me that the book's original title was *Amalur,* a Basque word meaning "Mother Earth."[82] To make the book more "enticing" to English-speaking readers, Oakes retitled it *Green Fire,* basing this choice not on Leopold's vision (of which he was aware), but on the Dylan Thomas poem "The Force That through the Green Fuse Drives the Flower."[83] Thomas's poem allies greenness with the combustible, while also reconciling life and death. Oakes's change of title was most apt. Photosynthesis *is* the green fuse, *is* green fire, *is* the green fountain—*Amalur* in action, making life.

FAKING GREEN FIRE

The green Venus, a.k.a. Mother Nature-Earth, was known in ancient times as "the power that stirs seeds to grow, mature, flower and reproduce themselves . . . that renews plant life in its season and imbues animals with the urge to mate."[84] Extra-keen perceivers like Hildegard and Dylan Thomas invented words—*viriditas* and the *green fuse*—to name this creative and creating power.

Other cultures though, have suitable words at the ready. In Anishinaabe (the language of her ancestors), Robin Wall Kimmerer finds *Puhpowee,* which translates as "the force which causes mushrooms to push up from the earth overnight."[85] Kimmerer grasps this word "as a talisman," because it bespeaks the reality of a "world of being, full of unseen energies that animate everything."[86]

Envious types who want to master and mimic Mother Nature-Earth have long sought to capture this force. Alchemists wanted green fire to be able to copy nature, while contemporary synthetic biologists and genetic engineers want not only to copy but to change nature according to their will. Curiously, green fire remains the sign of that effort. In 1994, biologist Martin Chalfie and his team inserted the fluorescence gene of a jellyfish into a worm with "glowing results": "The worm's nerve cells shone bright green." Subsequently, manipulation with GFP (green fluorescent protein) has emerged as "one of the hottest tools in genetics. Today, things glow that were never intended to glow: cancer cells, zebrafish, potatoes, bunnies."[87]

The inclusion of "bunnies" refers to an incident in 2000 when artist Eduardo Kac ventured into "transgenic art," which he defines as involving the "use of genetic engineering to transfer natural or synthetic genes to an organism, to create unique human beings."[88] The goal is a technologically assisted form of becoming a mother by creating a new being, but one dependent upon the frontier violation characteristic of paradigmatic motherfucking. Kac contracted with a lab in France to insert jellyfish EGFP (a synthetic version of GFP) into an albino rabbit embryo. When placed under a certain light, this rabbit reportedly glowed green (Figure 6.8). Kac dubbed her "Alba" and intended first to display and live with her in a gallery and then to "socially integrate her" by taking her home as a family pet. The laboratory ultimately refused to release her. Still, the response to Kac's project, titled *GFP Bunny* and memorialized by his photos of the glowing green rabbit, was enormous (Figure 6.9).[89]

The rabbit in Kac's artwork is real flesh and blood. She stands, along with her species, as "an ancient symbol of fertility," seen even today in the Easter bunny.[90] The rabbit as symbol communicates "that great mystery of life renewed through death,"[91] in other words, Mother Nature-Earth. Kac's rabbit was made to order in the lab. The ordeal to which he subjected her ritually reenacts *Man's* storied possession and manipulation of nature, while also allowing the artist to claim the mantle of creator. Kac continued to make art based on Alba's green glowing image. All that I can learn of Alba's fate is that she reportedly died prematurely in the lab where she was "made." I am left wondering: Had Kac been present at the moment of her death, what might he have glimpsed in her eyes?

FIGURE 6.8 Eduardo Kac, "GFP Bunny," 2000. Transgenic artwork. *Alba, the fluorescent bunny.*

Some critics lambasted Kac's use of a living creature and violation of her biological integrity. Others, including those who advocate transhumanism, applauded his intervention for disrupting what they deem outmoded and oppressive notions of integrity and fixed boundaries between species.[92] But the right to integrity,[93] to not be someone else's frontier, is precisely what human rights as well as Mother Earth Rights requires. Reflecting on his GFP Bunny project, Kac calls for responsibility in conducting transgenic art projects, which would include establishing a "dialogical relationship" with his subjects.[94] Certainly, there are precedents for this practice of consensual relationships. The renowned potter Corn Blossom (Margaret Tafoya) from Kha'p'oo Owinge (Santa Clara Pueblo) advises others working with clay to offer a

FIGURE 6.9 Eduardo Kac, "Nature" from the series *GFP Bunny—Paris Intervention* (2000), 22" x 34" (55.88 x 86.36 cm), dry ink on paper, open edition. Collection Verbeke Foundation, Belgium.

gift and directly ask permission before presuming to take matter from Earth: "You can't go to Mother Clay without the cornmeal and ask her permission to touch her. Talk to Mother Clay."[95]

The question remains: What is a culture to do if it has no traditions of communicating with other-than-human beings, let alone of asking permission? Ted Chiang's fiction "The Story of Your Life"—the basis for the film *The Arrival* (2016)—ponders this subject. Extraterrestrial visitors have abruptly landed in various places on Earth. World militaries fear that they have come to make war. A team of diverse experts is assembled to deal with the problem, including by learning how to communicate across species. A linguist sets out to learn the language and in doing so finds that she acquires, in addition to her normal "sequential mode of awareness," another "simultaneous mode." This enables her to perceive "all events at once" and to realize that there is "a purpose underlying them all. A minimizing, maximizing purpose."[96] The story's underlying ecological message is that humans need to learn the languages of other terrestrial beings—other than human and human—in order to attain an alternative green consciousness and end the war on life that Anthropocene *Man* is waging.

A related concern is persecuted human languages. The United Nations declared 2019 the International Year of Indigenous Languages, reporting that while there are five thousand different groups of Indigenous people, two thousand, six hundred and eighty Indigenous languages are now "in danger."[97] This is due to the continuing effects of genocide, colonization, and displacement.

Any loss of Indigenous languages is a loss for all. Robin Wall Kimmerer affirms that these tongues hold much value for stopping the ongoing "Age of Extinction" as they offer greening modes that might be adopted into conventional English.

Kimmerer hails the "grammar of animacy" found throughout Indigenous languages, which acknowledges the "life that pulses through all things" and "that lets us speak of what wells up all around us."[98] She deplores the use of the pronoun "it" in reference to "Grandmother Earth"[99] and advises using the "same words to address the living world as we use for our family. Because they are our family."[100] The Potawatomi tongue "does not divide the world into masculine and feminine," nor does the language speak of the "living world and the lifeless one" in the same way.[101] Inspired by an Anishinaabe word, Kimmerer suggests the singular *ki* and plural *kin* "to signify a being of the living Earth."[102]

All can take up a grammar of animacy, even those who do not speak an Indigenous language. All can learn and invent new lifeways and research protocols, ones that attend to the beingness and speech of plants, elements, and animals.[103] Even when an Indigenous language is lost, the Cheyenne elder Bill Tall Bull tells Kimmerer, there remains a basic way to communicate: "'If you speak it here,' he said, patting his chest, 'They will hear you'"[104] This is not a metaphor. Heart knowledge conjoins emotion, spirit, wit, and intuition and is integral to listening and speaking among the earthbound, human, and other than human.[105]

Nobel prize–winning cytologist Barbara McClintock practiced respectful communication with the beings, including corn, with whom she interacted in the course of her research. Evelyn Fox Keller details McClintock's appreciation of the vast ingenuity of nature and her requirement that scientists "listen to the material."[106] Keller describes McClintock's process as an "intimacy" with the subjects of her research, one "born of a lifetime of cultivated attentiveness" and serving as the "wellspring of her powers as a scientist."[107]

There now are calls for a scientific initiative on the scale of the government-sponsored Manhattan Project—which resulted in the atomic bomb—to address man-made climate change.[108] Such an endeavor to end *Man's* war on the planet would depend (as in Chiang's story) upon teams of diverse humans being able to achieve attentiveness and intimacy with each other and with other-than-human beings, including by learning their languages and sharing their consciousness. To do this with any chance of success, this project would need to reject knowledge systems based in human exceptionalism. They would also need to include a broad range of participants, including experts (such as Kimmerer herself) who respect grammars of animacy.

Another sort of specialist might be someone like Starhawk, who has developed expertise in European cunning traditions that were deemed heretical, punished, and discredited by patriarchal religion and science. These involve the ability to work with herbs, plants, and energies-spirits in the service of the bio-community. In an interview, Starhawk related a vision that she experienced in a trance undertaken with a friend who is a microbiologist:

> And I met the great microbial mother, because, really, you know, somewhere way back then, there was one microbe, one little string of molecules that linked itself up into a membrane and split apart and reproduced itself and life was born, right? . . . and she looked at me and she said, "I can take you out. If I wanted human beings off this earth, you would be gone in a minute and there would be not one thing that you could do about it. I want you to wake up and get it and get with the program and do what you're supposed to be doing. And if you do, then I will work with you. And climate change, you know genetic engineering, and pollution, all those things, we are the great planetary chemists. We could deal with that in a minute. But we won't until you wake up and start living in balance again. So we're not going to save you. But you start getting on the right path and you can call on us and we will be your powerful allies."[109]

Starhawk's experience correlates with traditions labeled witchcraft by the Catholic Church, as well as with Indigenous understanding of nature as intelligent, purposeful, and self-changing. Humans are *nature* too. Those of us now living in "imbalance" might modify ourselves to recognize that we are part of a larger community to whom we should be fulfilling our obligation to feed the green, to keep that ultimate home fire burning.

TENDING GREEN FIRE

Robin Wall Kimmerer's scientific specialty is mosses, and she tells a story about her transformative encounter with the green fire of a moss known as *Schistostega pennata* or Goblin's Gold. This moss grows in caves, where it catches its life-giving light only at sunset. It has no leaves or shoots, but only a "fragile mat of translucent green filaments," a "shimmering presence" that "glows in the dark."[110] While visiting a cave in the Adirondack Mountains, Kimmerer noticed that "something glitters. Something green. Something fleeting, like the eye of a bobcat in the firelight."[111]

Kimmerer does nothing so violent as kill a mother wolf, but she does have reason for regret. For some time after this first meeting, she gave respectful attention to the glittering green moss, returning often with students and friends. But at some point, distracted and focused wrongly on "the small tyranny of things," she neglected the moss. When Kimmerer finally goes back for a visit, she finds that the bank above the cave had collapsed under its own weight, closing off the mouth of the cave and allowing her to visit no more. To understand the spiritual meanings of this interaction, Kimmerer draws upon the tradition of the Onondaga, the people of that land, who understand nature as purposeful. Obviously, if people exploit other beings, those beings will leave one way or another. But Kimmerer explains that neglect can also have a similar effect: "Plants come to us when they are needed. If we show them respect by using them and appreciating their gifts, they will grow stronger. They will stay with us as long as they are respected. But if we forget about them, they will leave."[112]

Out of her remorse comes a new realization. The presence of fiery greenness is profoundly contingent. In this case, it depended on the coincidence of the cave's angle to the sun, the precisely right height of the hills on the western shore, and the direction of the winds that carved out the cave in the first place. This shimmering moss, first as presence and then as absence, teaches

Kimmerer that "its life and ours exist only because of a myriad of synchronicities that bring us to this particular place at this particular moment. In return for such a gift, the only sane response is to glitter in reply."[113] Continuous rebirth will go on with or without us. When humans glitter back, though, we renew the patterns that greenlight our continuance.

Aldo Leopold and most other ecological thinkers emphasize that, because evolutionary changes are characteristically slow, ecosystems have time to adjust. Anthropogenic changes, though, are often "of unprecedented violence, rapidity, and scope,"[114] and result in ecological disruption and a cessation of renewal—hence, sickened soils, contaminated waterways, loss of fertility, pandemics, famine, and erosion. Some shifts, like those resulting in climate change, are so profoundly troubling because, as physicist Helen Czerski emphasizes, it's not really about whether it is "warm or cold: it's about how the patterns change."[115]

Such patterns, what Kimmerer calls those "myriads of synchronicities" result in tides, seasons, gulf streams, animal migrations, etc. They are what have been keeping everything and everyone going until now. A group of Earth System scientists, addressing the dangers of the Anthropocene, focus on the ways that planetary patterns have come under increased "pressure" from "human impacts" or "forcings":

> Human impacts on the Earth System do not operate in separate, simple cause-effect responses. A single type of human-driven change triggers a large number of responses in the Earth System, which themselves reverberate or cascade through the system, often merging with patterns of natural variability . . . lead[ing] to further forcings that can alter the functioning of the Earth System.[116]

These authors recommend a stewardship of the planet based in an understanding of Earth as an integrated system. The goals they outline—avoiding ecological catastrophe, while also achieving sustainable lifeways and equity among members of the human population—make fine sense. Still, they are not enough. There also needs to be that understanding, as explained by Jack Forbes, of nature as intelligent, active, and self-modifying.[117]

Lee Maracle speaks of the "human responsibility to become familiar with that patterned behavior of the Earth, ally ourselves with these patterns, and augment our life within the context of Earth's patterns or suffer the consequences."[118] Leroy Little Bear further stresses humanity's obligation to know and actively work to renew the patterns that allow our existence:

> The Native American paradigm is comprised of and includes ideas of constant motion and flux, existence consisting of energy waves, interrelationships, all things being animate, space/place renewal, and all things being imbued with spirit . . . Renewal is an important aspect of the Native American paradigm . . . creation is a continuous process but certain regularities that are foundational to our continuing existence must be maintained and renewed. If these foundational patterns are not maintained and renewed, we will go the way of the dinosaurs. We will be consumed by the constant flux. Hence, the many renewal ceremonies in Native American societies.[119]

In other words, the humans who do not already do so need to find ways not only to stop their harmful forcings, but also to feed the green that sustains us. Already, on the horizon, a new pattern is asserting itself in the proliferation of mutagenic greens. These do not feed us. They might even eat us.

MUTAGENIC GREENS IN THE ANTHROPOCENE

When the *New York Times* reported that controversial biohacker Aaron Traywick had died accidentally, it included an illustration depicting him glowing a disturbing green.[120] Similar bright, garish, strange, synthetic, and downright mutagenic greens are used to identify transformed superheroes, strange rays from outer space, scary monsters, and toxic substances in much pop fiction and art. Whether these new greens are said to derive from lethal chemicals, nuclear radiation, or alien visitation, they symbolically announce that life endures in the Anthropocene, but not as we have known it.

Mary Shelley described her monster in *Frankenstein* as being a yellowish color.[121] But in the poster for the 1931 Universal Studios film, he appears eerily green-skinned, and that color has stuck. Universal Studios actually copyrighted that feature, along with the bolts in his neck.[122] The monster's troubled and troubling greenness signals not only the corpses Dr. Frankenstein used in the production of the monster, but also Shelley's core theme: the blowback disaster that ensues when men try to change nature's patterns, when they play Mother Nature.

Qualitatively transformative greens in popular culture are found, for example, in *The Green Slime, Teenage Mutant Ninja Turtles, Flubber, The Hulk, Lego Marvel Super Heroes, The Mask*, and *Little Shop of Horrors*. Depending on the genre, the resulting mutation can be tricky, tragic, or empowering, but the basic meaning of this weird greenness remains constant: Nothing will ever be the same again.

Dr. Seuss's *Bartholomew and the Oobleck*,[123] appeared in 1949, soon after the first atomic bomb explosion, and memorably featured a dangerous mutagenic green. An arrogant king tires of the predictable rain, sunshine, fog, and snow that fall from the sky, and demands that his magicians make something new. The "oobleck" they oblige him with is a falling, sticky, and grossly green substance, which takes over everything, burying even the magicians and rendering the king "helpless as a baby." The day is saved only through the intervention of a boy who advises the crying and frightened king to apologize and to admit that the crisis has been "all his fault." When the king does, the oobleck vanishes.

The story is not just a diagnosis, it also offers a remedy. The child insists that the king declare a new annual holiday, one that expresses undying gratitude to the rain, sunshine, fog, and snow. Saying "please," "thank you," and "I'm sorry" and reciprocating kindness and gifts is good etiquette, the very foundation of "environmental ethics"[124] for the interdependent community that *is* Mother Nature-Earth. When members of that community neither hoard nor trash elemental gifts,[125] this ensures—to the extent that is possible in an always changing world—that the (green) fountain of energy continues to include them in its rounds.

More mutagenic green makes its presence known in *Annihilation*, the first novel in Jeff Vandermeer's Southern Reach trilogy (as well as in the film *The Annihilation* (2018) based on it).[126] The story opens as some unknown radiant force abruptly strikes a lighthouse, initiating a foundational alteration in the DNA patterns of life on Earth that mixes up the DNA of humans, plants, and animals. Humans prove helpless to resist. Civilization retreats; a rainbow shimmer in the air blankets all, and a prodigious if threatening growth of the green ensues. When a mutated scientist dies of injuries, "a thin green fountain of light" gushes up from her wounds, suggesting "some sped-up form of life burning fiercely."[127] A now literate and glowing green fungus becomes, veritably, the handwriting on the wall, spelling out in cryptic sentences the foundational alteration that is happening.

In *Annihilation* humans are overwhelmed by a force far greater than themselves. A similar situation takes hold in Louise Erdrich's novel, *Future Home of the Living God* (2017). Evolution has started to run backward, and Erdrich's narrator muses, "I cannot imagine how everything around us and everything within us can be fixed. What is happening involves the invisible, the quanta of which we are created."[128] Possibly, she also muses, the root cause is that "God has decided that we are an idea not worth thinking about anymore."[129] In other words, the life source-force has shifted attention away from us; the underlying patterns are changing inexorably, and humans are not the ones in charge.

Theorist Donna Haraway also has mutation and altered DNA in the Anthropocene on her mind. In *Staying with the Trouble: Making Kin in the Chthulucene*, she proposes a thought experiment, "an ongoing speculative fabulation" involving, in part, the establishment of communities where people are free to "choose a gender—or not,"[130] everyone has at least three parents, and for some, a possible path entails "the right and obligation of the human person, of whatever gender, who is carrying a pregnancy to choose an animal symbiont for the new child."[131] This child, at birth, is given "a few genes and a few microorganisms from the animal symbiont,"[132] who is from a threatened species, for example, a monarch butterfly. The symbiont species is not subjected to modification and the human may choose to add more of the symbiont's genes in the future. The changed humans have particular affinities with their symbionts and are tasked with the lifelong stewardship of their partner species, "nurturing . . . and being nurtured in turn."[133] Haraway acknowledges the possibility of making ecological and political mistakes. I do so as well and in that spirit ask a few questions. How would this society ensure that the symbiont's genetic material is given consensually? What distinguishes this practice from that of extractivism? Does the choice to conjoin belong only to the human partner? Why does this transfer of DNA involve an infant rather than someone old enough to choose?

The gene-editing technique that would make the last part of Haraway's fabulation possible is known as CRISPR. The title of one article proclaims that CRISPR now "Paints the Future of Genetic Engineering"[134] and illustrates CRISPR using a mutagenic green color. The technique itself, derived from inventive bacteria, holds out the promise of ending the suffering caused by certain diseases and conditions. At the same time, it has marked eugenicist potential.

Biochemist Jennifer Doudna, one of the scientists who developed the CRISPR technique, describes it as "powerful biotechnology tools to tinker with DNA inside living cells," enabling scientists to "manipulate and rationally modify the genetic code that defines every species on the planet, including our own."[135] The plus side of CRISPR, in Doudna's eyes, is that "humans" now have "primary authority over life's genetic makeup" and will be able to supplant the "deaf, dumb, and blind system that has shaped genetic material on our planet for eons" with a "conscious, intentional system of human-directed evolution."[136] This, unfortunately, sounds a lot like Anthropocene *Man's* disastrous belief in his own superiority over supposedly mindless and passive nature.

Still, at night and in the dark, Doudna's dreams guide her to other considerations. In one nightmare, a tsunami approaches a place she once knew as home, Hilo, Hawai'i. Even worse, in another dream Hitler appears and asks her to teach him the CRISPR technique. Doudna recalls the nuclear scientists who created weapons of mass destruction with good intentions but lived to regret it. Unhappily, she "feel[s] a bit like Dr. Frankenstein" and wonders if she, too, has "created a monster?"[137]

Navajo or Diné people have much to say about nuclear technologies and monstrosity. They provide a spiritual interpretation of nuclearism, and name *uranium* as a monster of the worst sort. A good deal of uranium was extracted by the US government on Navajo lands using Navajo miners who were not informed about the terrible dangers involved. This resulted in their own disease and death as well as sickening effects in their families and through the generations. In their "Navajo cultural interpretation of uranium," Esther Yazzie-Lewis (Navajo) and Jim Zion explain that "the Navajo word for monster is *nayee*. The literal translation is 'that which gets in the way of a successful life.'"[138] To thwart a monster one must name it, thereby gaining "knowledge of its destructive force"—knowing what it can do to you as well as "what you can do to it."[139] *Naming* includes identifying networks of interconnected monstrosities, including the war industry for uranium.

Monsters, in Navajo diagnosis, emerge out of a context of "disrespect for both other humans and the environment." They "feed on power . . . Their nourishment comes from disrespect."[140] Yazzie-Lewis and Zion prescribe a refusal to give monsters what they need and instead institute solidarity and consent in all relations, the giving of "prayer and good thoughts" and practicing constant "respect" for "that which native societies keep dear: Mother Earth and the Five-Fingered People."[141] In keeping with that practice, some Indigenous groups perform rituals of apology and gift giving to Earth in the wake of nuclear depredations, including the atomic bombing of Japan and the meltdown of the Fukushima reactor.[142]

Doudna hopes to avert the potential monstrosity of CRISPR with public education, followed by a collective decision regarding its use. This is sensible, but incomplete. That education must criticize and fully reject a guiding vision based in human superiority over supposedly separate, passive, and unintelligent nature. Doudna concludes by calling for caution as scientists "unlock . . . nature's secrets."[143] She rightly demands responsibility. But responsibility requires the abandonment of the notion that nature holds "secrets," traditionally in her "womb," which scientists must heroically "unlock." This *unlocking* has long been a euphemism for rape, for breaking and entering without invitation, disrespecting the "Mutha'" and thus serving up the very stuff on which monsters feed.

The final word here goes to the mutagenic green and queer cult classic *Little Shop of Horrors*.[144] The story is all about a world turned upside down as plant-monsters feed on humans. The tale originates in a germinal short story from 1932 called "Green Thoughts."[145] The same story then became the basis for an exploitation film in 1960, then a hit Broadway musical in 1982 (written and directed by Howard Ashman), and finally in 1986, another film, this time of the Broadway musical. In all of these, the plants end up swallowing almost everyone. But after test audiences for the first version of the 1986 film rejected that apocalyptic ending, the studio changed it, tacking on a "happy" conservative ending in which the central heterosexual couple unites and survives and the plants are defeated. In 2012, however, the film's director, Frank Oz, released another version of the film, restoring the "unhappy" finale.[146]

Little Shop of Horrors (2012) opens with a sudden and unexpected total eclipse of the sun. A blast of eerie green lightning comes out of the black, zeroing in on an urban florist's street display and striking one plant. The plant mutates. Instead of photosynthesizing, it now feeds on humans. Although it is sustained for a while on the offered-up blood of the nerdy hero, the green plant continues to demand "feed me" and ultimately consumes all the main characters. The plant reproduces and the progeny grow into giants, greedily snacking on humans and broadly relishing the destruction of civilization by rampaging through cities a la another green monster, Godzilla.

The original plant climatically sings out: "I'm a mean green mother from outer space." The lyrics denounce those humans who have shown a poor grasp of etiquette, who remain totally oblivious to the forces that they have been "messin' with," and who must now face the consequences. The plant is voiced by Levi Stubbs, lead singer of the Motown band The Four Tops. He brings markedly Black speech to the role, criticized by some as minstrel-like, although Stubbs refutes this. The Black speech tones ensure, though, that the "mean green mother" is understood as a *muthafucka'* in that sense of the indomitable and inexorable force, the *"Mutha'"* who here encompasses Nature-Earth and Universe.

Little Shop of Horrors is mostly sociopolitically incoherent, but its unhappy ending does smack of subversion. The overall story speaks directly to a core spiritual meaning of the Anthropocene. The patterns that have allowed for our existence are changing in ways that signal the end of life as we know it. Green plants, giving fare and air, have allowed those green-based lifeways. But the monstrous mutagenic-green plant of the *Little Shop of Horrors* instead takes our breath away, consumes us, and destroys civilization, flipping (and then stomping all over) the script of human exceptionalism in which *Man* casts himself as the "eater of Others" who never himself "can be eaten."[147] Ironically, of course, from a *green* perspective, this demise of *The Man's* civilization actually *is* the happy ending. Bon appétit.

"WORD IS BORN"

BLACK ENGLISH CREATES THE AFFIRMATIONS "Word,"
"Word to the Mother!," and "Word is born,"[1] ones that are
rooted in the power of language. In the original edition of
Black Talk Smitherman notes that in "the African concept
of *Nommo*, the Word is believed to be the force of life itself.
To speak is to make something come into being."[2] Toni Cade
Bambara stirs in a bit of advice: "Words are to be taken seriously . . . Words set things in motion. I've seen them doing it.
Words set up atmospheres, electrical fields, charges. I've felt
them doing it. Words conjure. I try not to be careless about
what I utter, write, sing."[3]

Words are alive, *inspired*. In the act of utterance they are
filled with breath, with spirit, and hence are able to move both
speakers and hearers. I try to be careful while seeking inspiration from the genius of that charged word *motherfucker*. Take
note: *genius* carries dual meanings of "guardian spirit" and
"innate intellectual or creative power" (*OED*). *Motherfucker* has
both. There are many reasons for this, including the extremity of
conditions—abduction, rape, pain, chattel slavery—associated
with the word's emergence. Undoubtedly, too, the presence of
mother, perhaps the first and most charged of all words, matters
mightily.

As already noted, *motherfucker* has both negative and positive connotations. In this way, the word itself exemplifies the

Call Your "Mutha'". Jane Caputi, Oxford University Press (2020). © Oxford University Press.
DOI: 10.1093/oso/9780190902704.001.0001

audacious trickster tactic that Black Talk calls *flipping the script*. Smitherman elaborates:

> To reverse the meaning of unfavorable words and statements by recasting the "script" from your own perspective; to change the outcome . . . by reversing the terms of the argument. A kind of semantic inversion wherein "bad" becomes "good," and the tables are turned—in your favor.[4]

As *motherfucker* shape-shifted, the semantic became mantic, the magic was worked, and *motherfucker* transformed and now appears "in both negative and positive ways, and sometimes as just a filler with no meaning at all."[5]

As always, context is everything. When Janelle Monáe, in 2018, avowed herself to be a "free ass motherfucker,"[6] some likely recalled soul musician Curtis Mayfield, who in a 1977 lyric affirms his "plan to stay a Black motherfucker."[7] Poet and scholar Fred Moten, decades later, reiterates the vow in resistance to social anti-Blackness: "I do plan to stay a believer. This is to say, again like Mayfield, that I plan to stay a Black motherfucker."[8] Scholar Marlo D. David, in *Mama's Gun*, pays heed to vernacular African American culture, populated by "baaaad muthas"—like "a strong outspoken grandma, a sexually bold and attractive woman with no children, or even the fictional Black private detective Shaft."[9]

Others, too, gravitate to the magical power of the word. Writer and LGBTQ activist Michelle Tea, in 2013, founded an online magazine *Mutha*, "exploring real-life motherhood, from every angle, at every stage."[10] Artist Chris E. Vargas conceived of an "imaginary museum" of queer and trans history named MOTHA—Museum of Trans Hirstory and Art.[11] Playwright Minita Gandhi performs her own creation *Muthaland*—about her transformational survival of a rape attempt in India, the place of her birth.[12] Writer Cheryl Strayed, in her persona as an advice columnist, aptly recommends to one would-be-writer despairing of sexism: "So write . . . Not like a girl. Not like a boy. Write like a motherfucker."[13]

Motherfucker is so irresistible because the word holds swear powers—the especially concentrated potency to speak into being by invoking or calling the "Mutha'," the very source of the Word, the one birthing the Word. *Motherfucker's* flippancy, moreover, doesn't just upend. It ensures that you end up elsewhere, that you come out with a different outcome. The most comprehensive script flipping, political scientist Cathy Cohen explains, aims for a total "transformation of the basic fabric and hierarchies that allow oppression."[14] This is the cosmic "shift" Gloria Anzaldúa describes, as more come to the realization that they "share a category of identity" with all life and therefore start to "work actively to see that no harm comes to people, animals, ocean . . . to take up spiritual activism."[15] This conjoined work leads to the birth announcement of a world spoken in the "Mutha'" tongue that forbids oppression. It's not a promise of utopia—an impossibly pure, perfect, and changeless state—but a recognition that life on earth has been and still is better in different times and places, making real the possibility of being better again.

The basic Anthropocene script has it that *The Man* fucks Mother Nature-Earth and all those defined into nature. Feminist philosopher Nancy Hartsock deconstructs this narrative. "The man" or "the master" represents himself as an "omnipotent subject," putting himself "at the center," while constructing "marginal Others as sets of negative qualities."[16] The charged word *muthafucka'* inverts all that, getting rid of the omnipotent master, while also switching up the syntax. The story is now of the *Mother Who Fucks*.

In this reformulation, *fuck* becomes an intransitive verb—one that moves but takes no object, one that changes it up precisely because it doesn't objectify. The *Oxford English Dictionary* reports that the verb *fuck* "typically, esp. in early use" assumed "a man as the subject," with a woman or her equivalent as the object. Flip that, and the *"Mutha'"* reappears as that feckful, formidable, and fulsome ("sexually unrestrained, unchaste," *OED*) force she always has been. This "Mutha'" and her devotees have no need to negate, rape, or possess another. The "Mutha'" ceaselessly reminds all that rape is not an inevitable fact of life, and inspires those who demand the end of rape and mother-fucking in all manifestations.

The Mother Who Fucks, dare I say, *copulates*. Although not very erotic sounding, the Latinate word *copulate* means to "con-join" (*OED*), to unite what has been disastrously split—like up and down, light and dark, humans and other than humans. The Mother Who Fucks *is* a copula, a moving force that ties everyone together, the efficacious power characterizing Sex and Love goddesses like Aphrodite, Venus and the African diasporic Oshun.[17] This "Mutha'" is the "bridge" Audre Lorde talks about, the one "formed by the erotic," conjoining the sexual, the "spiritual and the political."[18]

The Anthropocene assumes "human" (*Man's*) separateness and superiority, including the notion that only humans have agency and only humans are aware. It is imperative to recognize the role of *Man* and his *Missy* in absolutely causing, for example, pollution because of greed and overconsumption, and to remediate whatever our individual, community, and national complicity might be. It also is of deepest importance to keep in mind the agency of nature. *Man* does disastrously contaminate the air with excess carbon dioxide and other gases wrenched out of their places, along with some man-made poisons. But climate change itself is the "Mutha's" response to the insult, a means of self-regulation. *Man* is destroying and disrupting ecosystems, a causative factor in mass extinction. But why not consider, as scholar and writer Paula Gunn Allen (Laguna Pueblo) once suggested to me, that it is the animals and plants who are abandoning us.[19] These considerations are good for your mind, essential for developing skills of recognizing and listening to the entire bio-community, while also hoping to get a Word from and with the force-source.

WORD FROM THE "MUTHA'"

Religious studies scholar Melanie Harris declares that "in an era of the Anthropocene," the "prophetic Spirit, which gives voice to the speaker, reader, student and scholar of ecowomanism . . . is an important aspect of the discourse."[20] The Earth "Mutha'" story, collectively issuing from the ground up, is this sort of inspired and inspiring one. As that story perceptively speaks to what's going on, it also participates in bringing it on, making it happen.[21]

Poet June Jordan also takes up this matter of inspired speech, concluding that "the creative spirit" as it moves through art, literature, and being is "nothing less than love made manifest." Boldly, she identifies that creative spirit with *mothering*, the most "important . . . difficult, and purely loving" task in all the world. Such mothering is of the deepest import, Jordan says, because "children are the ways that the world begins again and again."[22]

Jordan, a revolutionary, anti-racist, bisexual mother and public intellectual, makes this case while discussing the mighty importance of children's literature. She is clear that the *mothering* she is referring to is not something done only by the biological birth-giver, but includes the writers and illustrators who serve children by giving them the stories they need to grow. Nor is mothering confined to the essential (if unjustly made menial) tasks of making a home, feeding, listening, teaching, cleaning-up, and caretaking. To mother is to cohere all into community, to lead, story-tell, create, and empower self, child and group.[23]

Creativity is not purely individual, internal, or exceptional. Most artists know this, Jordan continues, for sometimes "an entire poem . . . and/or . . . a completely formulated, fictional character will be 'given' to us."[24] This gift from that "creative spirit" carries an obligation to get it out, to birth it into the world.

The anthropologist Deborah Bird Rose, reporting on a conference on the Anthropocene that took place in 2014 at the University of California, Santa Cruz, relates that participants kept coming around to the importance of creative storytelling as an agent of "transformation."[25] This is because, she elaborates elsewhere, "the stories we tell are powerful contributors to the becoming of our shared world."[26] So too, Donna Haraway argues for a turn away from the phallic or "prick tale of Humans in History . . . We must change the story; the story *must* change."[27] Part of that change is to recognize, as Serenella Iovino and Serpil Opperman do, that not only do stories matter, but matter is "storied," and this is not, they stress, a "metaphor."[28] The stories carrying the spiritual meanings and the creative chops to take on the Anthropocene are what Jordan speaks of as the ones that are "'given' to us." The truly transformative stories can't be top-down ones, and humans can't come up with them all on their own.

The English writer Harold Bayley, back in 1912, sought the meaning and sources for what he calls "wayside stories, familiar to uneducated people,"[29] and now mostly reserved for children's literature. This is the oral tradition, a source of enduring instruction and inspiration, which includes folk and fairy tales featuring powerful other-than-human beings—fairy godmothers, giants, talking trees, tricky witches, and animal helpers. It is true that some of these beings want to eat or thwart the human characters; others are willing to help those possessing a good heart.[30] The story of Cinderella is one such wayside story that has continuing meaning, including for the Anthropocene.

Cinderella starts out sitting in smoldering ashes, neglected, used, and abused. By story's end, with assistance from the fairy godmother (who appears often as a tree or a heifer), the darksome Cinderella emerges from those cinders and shines for all to see. Cinderella, it turns out, is the one who, even in the times that are most bereft of spirit, has been keeping "the fire alight."[31] The Cinderella story, Bayley tells us, is not about finding your prince, but about the survival of the *anima mundi*, the "Holy Spirit,"[32] the world soul, a.k.a. the "Goddess of the Earth."[33] The soul survives through even the most grievous oppressions. Matter and spirit rejoin when Cinderella, at last, emerges again to darkly shimmer in the glorious gowns from the godmother, which are reminiscent of the sparkling oceans and stars in the black sky. Bayley makes it clear that such subversive stories "have not descended from the educated to the uneducated classes, but *vice versa* have emanated, as it were, from the soil."[34]

Margaret Atwood tells the origins of her ecologically compelling novel *Oryx and Crake* in a manner that suggests the crakes themselves are the messengers if not progenitors of the story. Atwood was in Australia, relaxing after a book tour and visiting a bird reserve. She looked over a balcony and saw "red-necked crakes scuttling about in the underbrush." In that moment, she says, "*Oryx and Crake* appeared to me almost in its entirety."[35]

Zoologist Katy Payne, founder of the Elephant Listening Project at Cornell University, was one of the first to attend to the low-frequency rumblings of the elephants she encountered at a zoo. This was the beginning of her research that resulted in the "discovery of infrasonic communication in elephants."[36] Western ways of thinking herald Payne as the "discoverer," but Payne flips that, and also gives evidence for the efficacy of dreams as a way to knowledge:

> I don't know where dreams come from but on the night of the day that I realized we had discovered this immense amount of communication that no one had known about in elephants, I fell asleep and dreamed . . . that I was surrounded by elephants . . . and they were reaching out to me with their trunks, sniffing me the way elephants do. And then the matriarch of the group spoke . . . "We did not reveal this to you so that you would tell other people."

Many people reading this would assume that the elephant was instructing Payne to keep secret what she had learned. But Payne realizes something more subtle and meaningful in the elephant's words. She reports that she "woke knowing that the elephants had revealed it to me, not that I had discovered something. See, that was the message.[37]

Some cultures, though not Payne's and my own, do know something about where dreams come from. Some dreams can be particularly meaningful, carrying communications from ancestors and other than human beings, and thus are a "way of learning."[38] In Payne's dream, she comes to know that her research was not heroically individualistic, as Western ways would have it, but profoundly collaborative across species, with knowledge transmitted to her by the elephant matriarch.

Concourse, collaboration, kinship, trickery, as well as enduring partnerships among humans and other beings, are real—science, not superstition. Scholar Kim TallBear makes plain that she appreciates the ways that contemporary academics are coming around in such fields as "animal studies" and "new materialisms" to this recognition. Still, she says "indigenous peoples have never forgotten that nonhumans are agential beings engaged in social relations that profoundly shape human lives." TallBear is not talking only about clearly animate beings like animals or plants, for in some traditions: "'Objects' and 'forces' such as stones, thunder, or stars are known within our ontologies to be sentient and knowing persons."[39] Such persons think, speak, know, narrate, dream, and have their own agendas.

Humans are not the only ones with the word. The story capable of undoing the Anthropocene is the story *of* Earth as a choral community—a community with a common source and a common fate.

"WHO'S YOUR MOTHER"?

The popular taunt "Who's your Daddy?" is all about "who controls you."[40] It's mostly posed as a rhetorical question, "a boastful claim of dominance . . . sometimes . . . a derogatory claim of sexual dominance."[41] It carries, even if used in a "lighthearted" sexually teasing way, Juana María Rodríguez recognizes, hurtful implications "of rape and incest."[42] The phrase, like the verb *fuck*, fuses heteropatriarchal gender, sex, and violence as it encapsulates the traditional setup where father-master is overtly the protector-head but, at the same time, the most likely threat to his lover and family.[43]

In *The Sacred Hoop: Recovering the Feminine in American Indian Traditions*, Paula Gunn Allen writes that in the culture of Laguna Pueblo the most "important question" is "Who is your mother?" The *mother* in question is not only the individual who gave birth to you and raised you "or her equivalent," but the entire "generation of women" who participated in shaping you. Your "mother" also includes your "context," your "matrix."[44] Allen is writing about Indigenous cultures, but the question is pertinent to all the earthbound, as is her parting admonition: "Failure to know your mother . . . is failure to remember your significance, your reality, your right relationship to earth and society. It is the same as being lost—isolated, abandoned, self-estranged, and alienated from your own life."[45] "Who is your mother" is a question that is first of all personal; responses can be sometimes happy, sometimes painful, sometimes unanswerable. Mothers, too, can be terribly abusive, including sexually, and neglectful. Even when they are not, they are subjected to misogynist and often scapegoating stereotypes, more or less legal domestic rape, and church and governmental controls over their bodily integrity, including in the 21st century in the United States as state legislators increasingly favor the legal personhood of a fetus to take precedence over the personhood and human rights of the pregnant woman.[46]

"Daddy" culture draws a strict divide between "good" and "bad" mothers. The latter can be single and/or poor mothers in need of government assistance, mothers who are raising children with different fathers or "mothering while Black"[47] or brown, Native, immigrant, disabled, lesbian, trans, two spirit, queer, or in any way unconventionally gendered (and often many of these all together). Arbiters of moral, sexual, racial, and social "hygiene" have subjected those they deem "unfit" mothers to extensive controls over their sexual and reproductive lives.[48] They shame them for reproducing, subject them to mandatory use of dangerous contraceptives,[49] arrange for their children to be removed from them, [50] incarcerate them, forcibly sterilize them through medical procedures or through exposure to environmental toxins.[51] Being implicitly profiled by a racist society as an undesirable or "bad mother" is often fatal. For example, African American, Native American, and Alaska Native women die from pregnancy-related causes at a rate about three times higher than white women. At the same time, Black infants are twice as likely to die as White ones.[52]

Another dire reality is that mothers' bodies everywhere, but especially those of Indigenous women who live in severely militarily and industrially contaminated places, carry toxins in their breast milk, body fat, and placentas that sicken them and that also pass into to the fetus and nursing infants. Their bodies, Katsi Cook decries, have been made into a dump or "landfill."[53] Cook points to a remedy of *reproductive justice*, a platform first formulated by activist women of color in the early 1990s. Reproductive justice is defined by the activists and public intellectuals Loretta Ross and Ricki Solinger as including these principles: "(1) the right *not* to have a child; (2) the right to *have* a child; and (3) the right to *parent* children in safe and healthy environments." Also basic to reproductive justice is the establishment of "sexual autonomy and gender freedom for every human being."[54] If reproductive justice were to be achieved legally and culturally, environmental justice would have to be taken fully into account. Women would be free to regulate their own birthrates, bearing only the children they or someone else would be able to care for. Every neighborhood necessarily would be a safe and healthy one. It was the mothers of gender egalitarian Indigenous communities, as Lee Maracle makes clear,[55] who ensured that everyone had nourishing (neither tortured nor plastic) food. If reproductive justice were to prevail, everyone would have access to healthcare, and women's human rights would mandate available and safe contraception, abortion, and prenatal/postnatal care, including for those with disabilities.[56]

There would be water, land, and air free of pollutants, as well as equitable distribution of the "natural common goods" (Cáceres). In *The Man's* outrageously inequitable world, it is, ironically, this scenario that is deemed to be an impossible "fairy tale."

The Anthropocene has been implicit ever since some men first installed themselves as lord and master through seizing, using, abusing, enslaving, and commodifying women's sexuality and reproductive powers and then moving on to do the same to other peoples, other animals, elements, and land. The way out of the Anthropocene necessarily includes establishing conjoined reproductive-environmental justice. This means ending "Daddy's" definitional domination and ownership of all defined into nature and instantiating respect for mothers in all of the forms they take—female, femme, male, trans, and non-binary, biological parents or not, mothers both physical and metaphysical, personal, communal, and planetary.

MOTHERING POWERS

The *Oxford English Dictionary* lists a number of meanings for *mother*, including "the female parent of a human being"; "a female ancestor, *esp.* with reference to Eve"; "an effeminate homosexual man, *spec.* one who acts as a mentor to a younger man"; a euphemism for "motherfucker"; and "the source of something." *Mother* also means "the earth regarded as the source, nurturer, or sustainer of humanity . . . Nature regarded as a fundamental, *esp.* protecting or nurturing, force." These definitions, however rich, still draw back from naming the full reality of the mothering power as encompassing twinned life and death, creation and destruction. Mary Daly does not shy away from this, though, defining divinity as "form-destroying, form-creating, transforming power that makes all things new."[57]

This power to make, shape, shake up, break down, and make new pulses through the anthology *Revolutionary Mothering.*

Mai'a Williams, a co-editor of the volume, shares that the "book came from a vision I had of mamas who believe in themselves and their children, in the future and the ancestors so fiercely they will face down the ugly violence of the present time and time again."[58] Another co-editor, Alexis Pauline Gumbs, disputes a tendency in queer theory to dismiss the values of mothering. She advocates "revolutionary mothering," as developed in Black feminist theory and practice, as a necessary way of challenging violence and working toward a just future and making all things new:

> We say that mothering, especially the mothering of children in oppressed groups, and especially mothering to end war, to end capitalism, to end homophobia and to end patriarchy is a queer thing. And that is a good thing. That is a necessary thing. That is a crucial and dangerous thing to do. Those of us who nurture the lives of those children who are not supposed to exist, who are not supposed to grow up, who are revolutionary in their very beings are doing some of the most subversive work in the world . . . We know it from how fearfully institutions wield social norms and try to shut us down. We know it from how we are transforming the planet with our every messy step toward making life possible.[59]

Revolutionary mothering refuses the passive, private, and self-sacrificial maternal mode that Daddy culture extols while also enacting systemic outrages on mothers.[60] Revolutionary mothering means empowerment of self, of child, and of community, including the earthbound community, often against devastating opposition. Mothering means growing and preserving life and the conditions under which life flourishes, including mandating nourishing food, determining how large a human population can be without precipitating collapse, and working for environmental justice. Revolutionary mothering makes an ecologically viable world by imbuing children with what June Jordan calls "the creative spirit's" principles of respect, love, and reciprocal nurturance. Without these, Jordan warns, "we will certainly not survive."[61] Another radical intellectual and mother, Irene Lara, explicitly links revolutionary human mothering to the mother-powering of *Madre Tierra, Madre Naturaleza*. Lara calls for a spiritually-activist "mamihood," based in the ancient under-standing of "revered Mother Earth, as symbolized by Cihuacoatl ('Serpent Woman')"and associated pre-Columbian goddesses.[62]

I have never given birth nor raised a child and have never aspired to do so, but I do strive to make revolutionary mother-ing, as well as alternative kinship-connections among human and other than humans[63] a core part of my life, particularly through my teaching. This is nothing exceptional. There are manifold modes of other-mothering. For example, in Black and Latinx gay, trans, and femme cultures a person, often an elder, can be recognized as a *mother*. Performance artist and scholar E. Patrick Johnson explains, "To be hailed as 'mother' is to be held in high esteem and regarded with great respect."[64] These gifted individuals are nurturers, mentors, leaders, counselors, and mediators. Mothering in these ways, Johnson explains, they perform "a resistive act against hegemonic heterosexuality and homophobia" and become agents "of community building and communitas."[65]

Mothering takes on these and other earthy meanings for Pueblo peoples, where mothering is a core practice for the well-being of all. Scholars and siblings Tito Naranjo (Kha'p'o Owinge) and Rina Swentzell (Kha'p'o Owinge) explain as much regarding their Tewa culture:

> In the social, religious, and political structuring, the term *Gia*, or mother, is a term used pervasively to identify ideal behavior. *Gia* is a term used to address the earth. It is also used for the highest supernaturals who remain in the underworld, for males who are outstanding leaders, for strong community level females, and for biological mothers. The human quality of nurturing is valued so highly that the predominant beings of the Pueblos are the *Gias*,

the Mothers, in all categories of Tewa social, political and religious life. They are people who love and help.[66]

The night after typing out that passage, I dreamed of a president of the United States—a grandmotherly type, who was also at the same time a formidable leader. I don't see this as prophetic, just a wish reflecting my angst in the midst of the Trumpocene. Would that all those who dare to lead and make decisions affecting so many be *mothers* in the Tewa sense. Would, too, that the script be flipped so that more of those assigned male at birth and identifying as straight and cisgender expand their self-understandings and practice so that they, too, do the mundane necessary work of mothering—caring, cooking, feeding, listening, home- and community-making—now exclusively assigned to a subordinated, feminine gender role but without which no one could live.

EARTH IDENTITY POLITICS

Paula Gunn Allen affirms that the answer to the question "Who is your mother?" holds the key to one's "own identity."[67] That answer brings in a lineage all at once personal, communal, and land-based. Linda Tuhiwai Smith explains that for Indigenous peoples, it is understood that everyone's "genealogy can be traced back to an earth parent, usually glossed as an Earth Mother."[68] *Identity* is a curious word, with its own revealing flip side. It means, on the one hand, "the quality or condition of being the same" and, on the other, "the condition of being a single individual" (*OED*). This seeming paradox is actually the truth of being earthbound. Jeanette Armstrong writes of her Okanagan culture:

> Our word for people, for humanity, for human beings, is difficult to say without talking about connection to the land. When we say the Okanagan word for ourselves, we are actually saying, "the ones who are dream and land together." That is our original

identity. Before anything else, we are the living, dreaming Earth pieces. It's a second identification that means human . . . as separate from other things on the land.[69]

The ways of thinking and acting that have wrought the Anthropocene are cemented into individualism and alienation from Earth and the earthbound. "Prick tales" (Haraway), including pop-culture narratives, continually recite the story of Man's heroic deicide of the "Mutha'." In these stories, the protagonist is an initiate who becomes a man by separating from and vanquishing a "monster" who represents Mother Nature-Earth.[70] An endless stream of stories and artworks from antiquity until today, for example, tell of Perseus beheading the Gorgon Medusa, displaying her severed head and trampling on her body.[71] A related tale is Disney's *The Little Mermaid*, in which another indistinguishable prince (with the besotted Ariel's help) rams a ship into the womb of the sea-witch Ursula, an unforgettable character modeled on the legendary drag queen Divine.[72] All such stories portray the initiatory hero (and complicit heroine) as successfully cutting the apron strings, so to speak, of the mother. But those "strings" represent the truly umbilical dependency of all terrestrial life on the context that is *Madre Tierra, Madre Naturaleza.* Cut those threads and cease to be.

THE TREE OF LIFE

The interconnected identity of life on Earth that *is* Mother Nature-Earth is affirmed not only in that original naming of the planet as mother, but also in the related concept of a *Tree of Life*, archetypally associated with the "Great Mother," the "matrix and the power of the inexhaustible and fertilizing water she controls."[73] Charles Darwin married ancient knowledge to modern science when he brought in the Tree of Life to illustrate his theory of evolution.[74] The trunk indicates that everyone's origin is the same.

The crowning, leafy canopy expresses the unity and interconnectedness of life as an exuberant multiplicity. Distinctiveness comes in with a species' placement on the tree's limbs, which branch out in different directions; sometimes neighboring branches might merge back together or create something new. My colleague primatologist Kate Detwiler, who actually did encounter a new species, put it this way: "Nature is always experimenting."[75] In other words, Mother Nature is a scientist.

A revelatory work of activist art, *Mother Earth Tree of Life* by darkdaughta[76] (http://www.creativeresistance.org/mother-earth-tree-of-life/) places at the center a woman—squatting, naked, with teeth-bared, She labors to give birth to the blue-green Earth emerging from her vagina. This mother gives birth to herself and all terrestrial beings over and over again.

This Mother-Tree plants her feet on reddish-brown ground, the same color as her skin. From her right foot spring waters, rising up to touch the sun in her right hand. From her left foot grow plants, reaching to the crescent moon in her left hand. The balancing of the transpiring green flora and fertilizing blue fluids speak to the conjuncture of elemental powers of earth and water. David Suzuki and Wayne Grady explain that "even a single tree in the Amazon rain forest lifts hundreds of liters of water every day." The tree then transpires or gives off the vapors from those waters into the air, which then fall down as rain only to "be pulled back up again through the trees."[77] Trees provide wood for fire and oxygen for breath; their roots reach into the Earth, holding the soil. Without trees, humans and many other beings would not be here.

Darkdaughta's artwork bridges the sexual, reproductive, spiritual, and political, as it includes a multicolored banner whose demands include: "END WAR; FUCK GENDER; NO SURVEILLANCE STATE; END ABLEISM; THE PERSONAL IS POLITICAL; MY BODY MY CHOICE; END THE OCCUPATION; NO LEADERS NO FOLLOWERS; NATIVE AUTONOMY NOW." This is a revolutionary Earth "Mutha'"—inspiring and infusing interrelated movements.

A significant feature inscribed on the Earth Mother/Tree of Life's torso/trunk is a Sankofa heart. The Sankofa, from the Akan tradition, speaks to the necessity of reaching into the ancestral past to build the future.[78] Africana studies scholar Christel N. Temple cautions against regarding the Sankofa heart as simply a "flat," inanimate "symbol." Rather, it is a "communicator," energetically initiating an "interchange of ideas, knowledge, wisdom, and philosophy."[79] The Sankofa heart speaks to both Mai'a Williams's vision of the simultaneity of "the future and the ancestors" and Alexis Pauline Gumbs's affirmation of revolutionary "Mamas who unlearn domination by refusing to dominate their children, extended family and friends, community caregivers, radical childcare collectives, all of us breaking cycles of abuse by deciding what we want to replicate from the past and what we need urgently to transform, are m/othering ourselves."[80]

The Sankofa way fusing past and present guides people to pathways based on specific ancestries. For example, farmer and healer Amber Tamm Canty affirms that through ecologically based planting, growing, and harvesting, Black people like herself are "breaking ancestral trauma" stemming from when her forebears "were enslaved to work with the Earth." Thus, she says, Black peoples' current Earth work "heals us and our ancestors."[81]

The Sankofa way reaches back to the ancestry that all human beings share, a heritage with crucial implications for the future. Undoing the damages of the Anthropocene requires remembering humanity's non-heteropatriarchal and non-slaveholding past, and drawing upon this history while envisioning and building a more equitable future. Heteropatriarchy has not always existed and has never been universal and—because of its paradigmatic motherfucking—is not only abhorrent but unsustainable. Revolutionary mothering power moves through all those who participate in undoing heteropatriarchy, transforming patterns of domination and shaping ones of reciprocity, bringing into being other ways, other worlds, and other words.

SOUNDING LOUD, LOOKING LOW

A core intent of rape—activists, theorists, and survivors underscore—is to silence.[82] So, too, does motherfucking intend to silence Earth. Philosophers Lee Hester (Choctaw) and Jim Cheney explain that in the Western knowledge tradition, "the world is not permitted to speak on its own behalf. It merely answers questions posed by human culture and answers these questions, not in its own voice, but in a vocabulary, and according to an agenda, not its own."[83] Hearing the Word of the Earth "Mutha'" inspires a change of heart and a change of direction. Literary theorist Barbara Christian urges all to resist the normative silencing and demeaning of women (which extends into all those *Man* defines outside himself and into nature, even aspects of himself) by looking not only on "high," but also and always looking "low."[84]

Looking low is crucial in the Anthropocene. While "high" theory neglects to notice the rapism at the root of the era, the "low" theory—carried by obscene words, vulgar joking, and activist slogans—loudly declares that reality. To look low, moreover, is to encounter what is "down there." In popular parlance this means the genitals as well as the ground. Hildegard of Bingen knew that the "fertile Earth is symbolized by the sex organs, which display the power of generation as well as an indecent boldness."[85] The genitals, just like the ground, are "storied matter" (Iovino and Opperman) with much to say. They materially manifest cosmic powers of "creation and procreation,"[86] thought, love, attraction, intelligence, connection, ecstasy, periodicity, feeling, fragrance, funk, and boldness. Without that "indecent boldness"—in a word, *impudicity*—there would be no generativity; without that grounded intelligence—that *dirty-mind*—there would be no conception, in all senses of the word.

Impudence is derived from the Latin *pudendum*, which means *cunt* (as it also does in English). *Pudendum* derives from *pudere*, to feel shame, and participates in the misogyny and sex negativity that motivates more obvious slurs like *dumb cunt*. Another insult *cunty* means something "suggestive of, or associated with the female genitals" and hence "despicable; highly unpleasant; extremely annoying" (*OED*).[87]

In the academe, the cunt is tangibly taboo. Somehow, though, it's long been fine to focus on the phallus. Even some feminist theorists often pay attention more to the phallus than to any exploration of female and femme signifiers. Why not, though, shift the terms of the argument, taking to heart Christian's powerful reminder that Black feminist traditions advise us to "look low, lest we devalue women in the world . . . [and] our voices no longer sound like women's voices to anyone."[88]

There is no feminine equivalent of the omnipotent phallus in the heteropatriarchal imagination. Psychoanalytic feminist Jessica Benjamin writes that female "representation at the same level as the phallus" is an impossibility because of men's fearfulness associated with the mother and because most everyone's psyches are already preoccupied by and with the phallus. As she sees it, because "representation of the body is organized and dominated by the phallus, woman's body necessarily has become the *object* of the phallus."[89] And the phallus is always the "doer," except, Benjamin notes, when the phallus is "done to" in castration anxiety.

Those of us of any sexual identity or gender expression who are motivated to flip the script on phallic "doing," a.k.a. motherfucking, must gird our loins, gathering impudicity. Flipping the script does not entail an intent to castrate or violate, which is just to mimic the oppressor. Rather, the bold move is to renounce misplaced fear of the mother, while aspiring to be like her, to be *cunty*—to speak out and look low. It helps to recall that the cunt, in traditions ancient and new, sacred and profane, is known as a mouth—and a bold and wordy one at that. In a 2004 essay,[90] I took up the implications of an insult hurled by art critic Maureen Mullarkey at the feminist writer Andrea Dworkin. Mullarkey was aghast at Dworkin's book *Intercourse*, which analyzes heteropatriarchal intercourse as ritually enacting men's contempt for women.[91] Mullarkey accused Dworkin of specializing in "cunt-speak."[92] Reading this, I knew in my gut, even in my "down there," that this term was most flip-worthy—instantly reclaimable as both praise and fighting word.

Mullarkey's aspersion might seem inventive, but it is merely a misogynist cliché. For example, in 2001 some male members of the French Parliament, rancorous about women's membership in what they wanted to be a purely male governmental body, made sneering remarks about "vaginal verbosity," invectives meant to silence a colleague as she stood to make a speech.[93] The electoral violence directed against Hilary Clinton in the 2008 and 2016 presidential races openly called her a "cunt" on posters and T-shirts, and slurred any women who favored her candidacy as voting with their "vaginas."[94] The English historian and feminist television host Mary Beard reported that she had received in the mail "a photo of her face superimposed onto a picture of female genitalia."[95] She knew that it was an attempt to shame her into silence. Scholar and public intellectual Brittney Cooper reports that a "senior scholar" deliberately tried to deride "the seriousness of the work" produced by a "collective of Black feminist women that I write and panel with about the politics of sex and pleasure." This senior scholar accused the collective of "writing from our pussies."[96]

An explicit and related representation of misogynist contempt surfaces in *Hustler*, illustrating Dworkin's argument about the ways that men use intercourse to subordinate and silence women. Under a caption reading "Lip service" is a photograph of a woman's face, in which a cunt replaces her mouth. The text below the picture reads:

> There are those who say that illogic is the native tongue of anything with tits. . . . It comes natural to many broads; just like rolling in shit is natural for dogs. . . . They speak not from the heart but from the gash, and chances are that at least once a month your chick will stop you dead in your tracks with a masterpiece

of cunt rhetoric. . . . The one surefire way to stop those feminine lips from driving you crazy is to put something between them—like your cock, for instance.[97]

Hustler is openly outrageous, but it really just says what everyone knows. A rape culture prescribes men's routine sexual assault of the women they live with. This is intended to shut the women up and effectively shut down any resistance.

Making *cuntspeak* into an insult reverses original honorific associations based in perceptions of Earth sounds. The most famous oracle of ancient Greece was at Delphi. The place name Delphi is often translated euphemistically as *navel* or *belly*, but as religious historian Mircea Eliade politely puts it, *Delphi* actually means "the female generative organ."[98] At Delphi, oracles were lowered into an underground passage so that that they might receive messages carried by the sounds issuing from the Earth's genitals. Other seers were known as "belly talkers."[99]

The chthonian Greek Medusa is known for her snaky hair, fangs, protruding tongue, eye-biting gaze, occasional beard, and shrieking voice (Figure 7.1). Originally a goddess, she was turned by patriarchal myth into a monster. *Gorgon* is from an Indo-European root *garj*, denoting a fearful shriek, roar, or shout.[100] Medusa's epithets include "Gorgon voiced." We might not be surprised to find that her face represents, according to Sigmund Freud[101] and other observers, the "maternal throat/vagina."[102] No wonder Perseus acted to silence her. The dread "cuntface" of the Medusa is the unmasked face and speaking mouth of the Earth "Mutha'." A key feature of Medusa's legendary monstrosity is that her formidable gaze inexorably turns men to stone. Stones, though, are alive and bespeak the Earth's endurance, past and future. To turn to living stone is to become one with the living Earth. The pagan thinker Glenys Livingstone perceptively asks if the collective comprehension of Medusa's "visage" understood her face not as horrific

FIGURE 7.1 Terracotta stand, ca. 570 B.C.

Gorgoneion (Gorgon's face)

Signed by Ergotimos

Signed by Ergotimos as potter

Signed by Kleitias as painter. 570 B.C.

Terracotta; black-figure, H. 2 1/4 in., diameter 3 9/16 in.

Photograph Courtesy of the Metropolitan Museum of Art, New York, Fletcher Fund, 1931

but actually "as one of Divine Wisdom," would we be able to "change our minds sufficiently so as to affect the way we relate to Earth, to being?"[103]

A goddess's nakedness, including leg-spreading, is in non-patriarchal iconography a sign of potency. The posture was a magical-sacred configuration long before it was rendered pornographic in order to disempower it.[104] A recognition of genital magic underlies rituals still performed by elder "Mothers" in

West Africa, in times of greatest social calamity. Religious studies scholar Laura S. Grillo explains that

> "The Mothers" are *postmenopausal* women who, having sur-
> passed the defining stage of sexual reproduction, are ambig-
> uously gendered. Like primordial beings, their incarnate
> power resides in that gender doubleness . . . The seat of their
> power is not only the womb, but also the vulva. Appealing
> to their sex as a living altar, the women ritually deploy their
> genital power to elicit the most perilous of curses as an act of
> "spiritual combat" against malevolent forces that threaten the
> community.[105]

Perhaps this ambiguously gendered, vulva- and Mother-identified cursing power came to the Americas on the lips and in the hearts of those held against their will on the slave ships, a power that wended its way into that still resonating curse and swear word *motherfucker.*

Women's "genital powers," including powers of speech, continue to make themselves known. Visual and performance artist Carolee Schneemann tells of an inspirational dream advising her: "WHY DON'T YOU LET VULVA DO THE TALKING?"[106] Chicana feminists boldly name themselves *panochonas*, from *panocha*, "contemporary slang for vagina." A *panochona* represents, scholar and writer Emma Pérez explains, "a formidable, impressive, woman, whether lesbian or straight."[107] Feminist storyteller and theorist Clarissa Pinkola Estés avers that to speak from the *panocha* is to access "the *primae materia*, the most basic, most honest level of truth—the vital *os* . . . the mother lode, the deep mine, literally the depths."[108]

The Earth, in the oldest human reckoning, *is* a primal womb. Cuntspeak then is the "Mutha'" tongue, a language that is the birthright of all the earthbound, one essential to flipping the script of the Anthropocene and ending up somewhere else. Fluency in that tongue is neither limited to nor guaranteed for those with a vulva.

Consider the *os* to which Estés refers. The *os* is "the opening of the cervix into the vagina" (*OED*), the gateway through which fertilizing sperm enters and through which all pass in a vaginal birth. The "hint" (in Lame Deer's sense of natural symbols that speak to the otherwise hidden nature of reality) that the *os* gives turns on the potency of liminal spaces, passageways between realms. Not surprisingly, the signifying *os* is a property of the penis as well. Annie Sprinkle, the ecosexual activist-artist, once gave a show called the "Public Cervix Announcement." Inserting a speculum, she invited her audience to view her cervix, and some of those viewers exclaimed that the cervix with the *os* at its center looked just like the head of a penis, which indeed it does.[109] Similarly, a large clitoris and a small penis can be indistinguishable. Many people are born with bodily characteristics of both males and females, revealing the limited explanatory utility of those binary categories.[110]

A cosmic *os* appears in Carol Prusa's *Nebula*, painted in silverpoint on a rounded dome. A central cluster of far but luminous stars (Figures 7.2 and 7.3) appears within a great black round. Encircling that round are interlocking ellipses that overlie forms suggesting fecundity and transformation. Recall that alternative use of *motherfucker* as neither positive nor negative, but simply a "filler with no meaning at all" (Smitherman). Prusa's *Nebula* hearkens us to a recognition of this underlying aspect of the "Mutha'," the nebulous filler that is the very stuff of the Universe—the cosmic context,[111] the flux, the matrix,[112] or "Background,"[113] the potential space from which/whom meaning emerges. Another word for this space/place is provided by the artist and singer Erykah Badu—a "wombiverse." Marlo D. David interprets Badu's word-work as a way to understand the womb "not as a bounded organ in the female body, but . . . an infinite source of creation."[114] *Nebula*'s announcement of the simultaneously dark and shining wombiversal "Mutha'" resounds in the star fields emanating from Amir Khadar's *Goddexx* as well. These artworks bring us face to face with the "infinity" whom Hurston says humans petition in changing times when new knowledge is needed, times precisely like the Anthropocene.

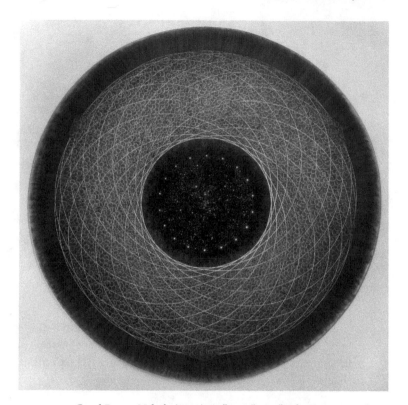

FIGURE 7.2 Carol Prusa, *Nebula* (2019), 60" x 60" x 10", Silverpoint, graphite, acrylic on acrylic dome with internal light. Used with the permission of the artist.

THE PRINCIPLE

A telling moment from the time of slavery is related by legal theorist Dorothy Roberts in *Killing the Black Body: Race, Reproduction, and the Meaning of Liberty*:

> A common recollection of former slaves was the sight of a woman, often the reporter's mother, being beaten for defying her master's sexual advances. . . . Minnie Folkes remembered watching her mother being flogged by her overseer when she

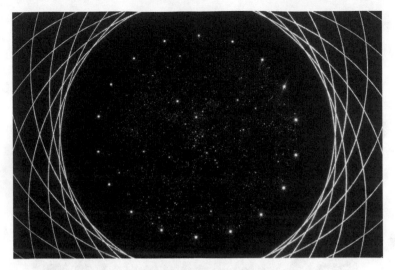

FIGURE 7.3 Carol Prusa, detail from *Nebula*, 2019.

refused "to be wife to dis man." Decades after her emanci-
pation, Minnie repeated with pride her mother's teaching: "Don't let nobody bother yo principle; 'cause dat wuz all
yo' had."[115]

The mother's use of *principle* carries strong philosophical-spiritual
meanings. *Principle* means "origin, source . . . source of action . . . a
primary element, force, or law; the ultimate basis upon which the
existence of something depends; Fundamental truth or law; motive
force" (*OED)*. The *principle* manifesting in the cunt is the "fundamental truth" that all the earthbound are obligated to respect the
inner space–outer space—where we came from and will return to.
The script for the motherfucker culture is all about violating the
principle, wherever that principle is found.

In 2017, the Women's March on Washington famously featured participants wearing pink "pussy caps," directly rebuking

Donald Trump, who had been videotaped boasting about his right to sexually assault women, including to "grab" them by the "pussy." Brittney Cooper, in her wittily titled "Pussy Don't Fail Me Now," reflects on the racial politics of the march and the caps, associated mostly with White and cisgender feminists. Cooper affirms the need to avoid centering White, middle-class, and cisgender women as the face of feminism, but she does favor the genital focus. Affirming that "black pussies matter," Cooper points to the crushing history of men's sexual violation of Black women, whose "vaginas were the property of plantation owners upon our arrival . . . used as a vehicle through which to reproduce plantation slavery." Furthermore, Cooper argues, current rates of rape indelibly show that genitals are political: "Having a vagina has always made us *more* rather than *less* susceptible to violence. Feminism taught me this. As long as that remains true, vaginas have to remain central to our organizing and our thinking."[116]

"Vaginas," in just this way, were central to M. NourbeSe Philip in her creative essay "Dis Place" (discussed in Chapter 2). Philip exposes the rape and maternal exploitation of Black women as the "fulcrum" of the plantation system and considers how rape also functioned as a means of silencing the "inner space" of the woman. Philip seeks "the sound of /the space/between/the legs/once found/how does the inner space/sound." She cries out, wanting to know "what does it say—this inner space . . .?" The silencing has been so total as to occasion the inner space to

> collapse in upon itself—/a black (w)hole absorbing everything around it.

Because, Philip says, no existing language can make that "inner space . . . whole," she proclaims the need for "a new language—the language of jamettes, possessing their inner and outer space. The be-coming and coming-to-be of a jamette poet."[117]

A *jamette*, Philip explains, is "a 'loose' woman, a woman of loose morals, whose habitat is the street. Jamette! A woman possessing both the space between her legs and the space around her. Knowing her place." With that self-possession, the "inner space," as Philip hears it, is "loud /like a jamette!/turning in/out—/that inner sound/found/loud/like a jamette!" Philip relates the jamettes to the Jamaican maroon freedom fighter Nanny, the historic warrior women of Africa, and the Mothers who exercise their genital power. All are quintessential *baaaad muthas.* Jamettes are akin as well, Philip says, to the "joying" women at Carnival, carrying their staffs or "linga" (penis symbols), flaunting their "yoni" (cunts) and "showing off their sexuality without undue censure or fear." It remains true that, still, "JAMETTES GET RAPED TOO!"[118] But their resistance is ongoing and their brazen actions fuel change as their impudent speech conjures another world.

"Jamette" speech is "loud," "filled with 'inherent potentiality'" and a quality of "polyvocality – the many-voiced one of one voice."[119] Cuntspeak is just such a many-voiced, thunderous and proud vulgate, issuing a deliberately dirty-minded rebuke--like the one powering a poster by activist-artist Favianna Rodriguez. The work has at its center the blue-green, round Earth (Figure 7.4), shown clearly as cunt in warm hues of red, orange, and yellow. Lightning bolts flash to either side of the labia, signifying revelation, sound, creation, destruction, and fertility, as well as the male energy intrinsic to the whole bodacious bundle.[120] This Earth-Cunt demands: "RESPECT WHERE YOU COME FROM." The text along the bottom makes it specific: "DON'T FRACK WITH ME."

Rodriguez explains: "The way we treat women is the way we treat the planet, and vice versa."[121] Rodriguez doesn't seem worried about being criticized for "reducing" women to a sex organ.[122] That accusation itself needs some serious flipping, as it is first of all heteropatriarchal dualism that "reduces" all sex organs by denying their intelligence, spirit, and voice. Rodriguez's Earth Cunt lays down the principle of *respect*, a concept that deserves elaboration.

FIGURE 7.4 Favianna Rodriguez, *Don't Frack With Me*. Copyright 2018 Favianna Rodriguez. Favianna.com.

Historian and philosopher Vine Deloria, Jr. teaches that because we are enmeshed in a "living universe," there is a firm requirement for "mutual respect among its members, and this suggests that a strong sense of individual identity and self is a dominant characteristic of the world as we know it. The willingness of entities to allow others to fulfill themselves, and the refusal of any entity to intrude thoughtlessly on another, must be the operative principle of this universe."[123] When humans act disrespectfully, for example, by mistreating other animals in factory farming, laboratory research, or by entering without care into wild animals' homeplaces to hunt, harvest, mine, or log, this has consequences, as when a hungry and opportunistic virus can jump to humans—perhaps the origin of the COVID-19 pandemic (unfolding just as this book goes to press).

This necessary understanding of *respect* is, not the usual "Who's your Daddy"-type demand for deference. Rather, this respect is based in strong and sure self- and other-awareness, even in the inevitable face of differences—and the concomitant requirement to act accordingly.

Some might find Rodriguez's cunt-speaking poster demanding respect to be problematically essentialist. But consider this wisdom from Brittney Cooper:

> Feminism wouldn't be feminism if it didn't celebrate the power of pussy. It might be trite but it's true. To be clear and not transphobic, feminism is also not about the elevation of particular body parts. My transwomen comrades have taught me that you don't have to have a vagina to have a pussy. And my lesbian homegirls have extolled to me the virtues of the "D" their own lesbian partners are throwing down.[124]

Pride in the *principle* enables the withdrawal of comfort and support from the heteropatriarchy, affirms feminist outspokenness and righteous rage, and insists on a feminism that is "first and foremost, about truly, deeply, and unapologetically loving women."[125] Cooper concludes her potent piece, "Pussy Don't

Fail Me Now," by flipping the script on that "senior scholar" who derided her group as "writing from our pussies." She cites this apt response from Black feminist theorist Michele Wallace: "What's wrong with writing [or organizing] from your pussy? It's a real place."[126]

Hear. Here. Cunts, pussies, or principles as you prefer—personal, planetary, and universal—are not real estate, but real places, with powers of place, powers of thought, and powers of speech. M. NourbeSe Philip knows that *The Man's* fear of jamette speech derives from this language's capacity for "signifying another reality,"[127] of ending this world and activating the power of *Nommo* to speak another into being.

CHANGING THE TERMS

Cunctipotence[128] is an old word, meaning "all-powerful" (*OED*). It derives from the Latin *cunctus* (all) and *potens* (powerful), but is now out of use; instead, *omnipotence* has become the word of choice. But the two are by no means synonymous. *Potency* itself is supposedly a neutral term, yet it's inseparable from an everyday meaning of a man's ability to get and maintain an erection. That notion of *potency* carries a great deal of classic misogynist baggage. St. Augustine of Hippo railed against Eve's sin that rendered men unable to control their erections at will.[129] Aristotle deemed woman to be as "it were an impotent male, for it is through a certain incapacity that the female is female."[130] Actually, Aristotle has it backward. In truth, it is through a socially imposed incapacity ascribed to women that the male is made male—that fictitious "100 percent"–type male Tomson Highway disavows. Take even one point off that perfect score, and a man or boy is derided as a "wimp," a "pussy," and even a "cunt."[131]

In *The Man's* world, everyone must be a woman or a man. For those assigned male at birth, simply having a penis does not make you a "man." Rather, a "real man" must seem to possess the fantastic phallus, which, as already noted, is nothing like the real penis. Stepping outside of phallocentrism, though, might allow the recognition of the penis as another kind of cunt or, at least, *not* a phallus. Theologian James Nelson finds in the penis a storied-matter manifestation of sacred *"relational* power," as well as a "generative power, the power of an open energy system."[132] Another theologian, elias farajajé-jones, wonderfully chants against *Man's* core divisiveness in his sexual-spiritual manifesto "Holy Fuck": "I'll be clit, vagina, vulva, labia. And I'll be dick and nuts and I'll be whatevah. I'll be 4-legged and I'll be 2 legged and I'll still fuck with your supremacy . . . "[133] This not only strongly reclaims the self, but simultaneously respects and powerfully invokes the communal, the "Mutha'," inner and outer.

In *Man's* reckoning, *power* is *omnipotence*, which is understood as stand-alone, unparalleled power over all. The *Oxford English Dictionary* notes that the word refers specifically to the "all-powerfulness of God," while *Webster's* offers the definition "unlimited authority." Here again is that *Big Baby* fantasy of separating from and then lording it over the mother and everyone else, respecting no limits in the quest for domination. *Omnipotence* is expressed through rape and rapism, slaveholding, authoritarianism, war-making, seizure of the common goods, hoarding of wealth and influence, and the demand for immortality. *Omnipotence* is the disturbed and disturbing fantasy behind the Anthropocene, where *Man* claims total power to rival and overwhelm (Mother) Nature, while, absurdly, denying his dependency.[134]

I came upon the alternative word *cunctipotent* first when reading the entry for *cunt* in Barbara Walker's *Encyclopedia of Women's Myths and Secrets.* Walker traces the origins of *cunt* to the "Great Goddess Cunti, or Kunda, the Yoni of the Uni-verse" (reminiscent of Badu's *wombiverse*). Then, following geographer and archeologist Michael Dames, Walker links *cunctipotent* etymologically to *cunt* as well as *country, kin, kind, cunning,* and *ken.* She defines *cunctipotent* as "all powerful (i.e. having cunt-magic)."[135] Most linguists would dispute Dames's and Walker's suggestions, but I take them seriously as folk etymologies that allow us to understand *cunctipotent* as naming an utterly different notion of *power*. Cunctipotence is *not* omnipotence. It is not about power *over* all. Rather, cunctipotence is the conjoining power *of* the all.

A 1988 illustration from *Artistes Indigenos Graphics* depicts three mothers (one African, one Indigenous to the Americas, and one a pre-Columbian goddess). The two humans hold corn in one hand, and hold up the Earth with the other. The goddess rests on the ground between them, nursing a child. A statement curving over the top of the planet reads "Mothers & Others Connecting All."[136] This is the copulative conjoining effectivity of the "Mutha'." The word *conjoin* hearkens back to *yoni*, the Sanskrit word meaning "womb, vulva, vagina; place of birth, source, origin, spring; abode, home."[137] *Yoni, join, juncture, yoke*, and *yoga* all derive from the Sanskrit *yu*, which means to unite. *Mother Earth* means the innate connectedness and kinship of the earthbound. Full awareness and experience of this state needs a name in English. I suggest *cunctipotence*, illustrated in the following three examples.

Farmer Amber Tamm Canty's specific life experiences have led her to "weave together the interdisciplinary ways of working & healing with earth in full spectrum." A brief biographical statement on her website tells us that Tamm Canty grew up in a family that had experienced five generations of city living, but from earliest memory she yearned for immersion in non-human nature. Then "*her father murdered her mother* (emphasis in the original)." Tamm Canty's response was to retreat into silence, and she ultimately found that this "*time in silence helped her realize that when she layed her mother's body in the earth, the earth literally became her mother* (emphasis in the original)." In that most painful and ritually charged moment, it seems that Tamm Canty was able to directly experience the process of continuous rebirth that *is* nature. She finds an answer to the enduring question "Who is your mother?" in knowing-feeling her identity with Earth.

Changed by this moment of cunctipotence, Tamm Canty commits herself to the "green and growing world, working with plants and making it her life's work to "bring people to their mothers; either by helping them connect with the earth or by telling them [her] story." She also affirms her conviction that "Momma Earth would always guide + support her."[138]

Another moment of cunctipotence occurs in Alice Walker's *The Color Purple*. The bold and fulsome Shug, a brazenly autonomous blues singer, is speaking to her lover Celie, after Celie has been able to begin to free herself from lifelong patriarchal abuse. Shug is speaking of spirituality, particularly her rejection of the conventional patriarchal "God" ("the old white man"), which has allowed her to experience an erotic conjunction with all being:

> My first step from the old white man was trees. Then air. Then birds. Then other people. But one day when I was sitting quiet and feeling like a motherless child, which I was, it come to me: that feeling of being part of everything, not separate at all. I knew that if I cut a tree, my arm would bleed. And I laughed and cried and I run all around the house. I knew just what it was. In fact, when it happen, you can't miss it. It sort of like you know what, she say, grinning and rubbing high up on my thigh.[139]

Shug knows it when she feels it. Like Tamm Canty, she goes from motherless to *motherful* in a moment of being.[140] The sexual-spiritual earthbound state she enters fully upends *Man's* self-absorbed story of an omnipotent subject, defined by separation from and domination of nature. Flipping that script, Shug lands in a *real place*, one of cunctipotence, composing and composed of the power of *all*.

The English writer Virginia Woolf describes still another such moment of unmediated being:

> I was looking at a plant with a spread of leaves; and it seemed suddenly plain that the flower itself was a part of the earth; that

a ring enclosed what was the flower; and that was the real flower; part earth; part flower ... The whole world is a work of art ... we are parts of the work of art ... But there is no Shakespeare, there is no Beethoven; certainly and emphatically there is no God; we are the words; we are the music; we are the thing itself.[141]

Tamm Canty's, Walker's, and Woolf's perceptions all affirm the state of cunctipotence, which includes accepting and returning the gift of the mothering power, feeding the green, experiencing any violation of another as harm to oneself, and awareness of the interrelatedness that is life. This experience *is* the call of the "Mutha'." In response, we can offer the full-throated affirmations—*Word; Word is born; Word to the Mother*!

CALL (ON) YOUR "MUTHA'"

"IN THE TRADITIONAL, BELIEF SYSTEMS of native people," Linda Hogan writes, "the terrestrial call is the voice of God, or of gods, the creative power that lives on earth, inside earth, in turtle, stone, and tree. Knowledge comes from, and is shaped by, observations and knowledge of the natural world and natural cycles."[1] In Anthropocene culture, though, this knowledge is spurned, and the terrestrial call is not returned.

The phrase "call your mother" is common enough. On the surface, it pokes fun at (stereotypically) needy mothers and sympathizes with (understandably) neglectful children. Down where it counts, though, "call your mother" is an age-old imperative for the earthbound, a prescription that, if ignored, threatens to sever the most important relationship of all. Why do some of us dwelling on earth, but accelerating toward "Globopolis," remain unaware of this? Val Plumwood gets to the heart of the problem: "Modernist liberal individualism teaches us that we own our lives and bodies . . . As hyper-individuals, we owe nothing to nobody, not to our mothers, let alone to any nebulous earth community."[2] So certain are some of us that we are outside and superior to nature that we proceed as if we can turn our back on Nature-Earth with no consequence to ourselves. The mothering power is owed nothing, not even a phone call.

Linda Hogan, Emma Goldman, Max Baginski, and Berta Cáceres all stress that (Mother) Nature-Earth calls to us and that this call requires a call back, a reckoning, an apology, a thank you, a return of energy, a ritual renewal of the relationship, even a visit

Call Your "Mutha'". Jane Caputi, Oxford University Press (2020). © Oxford University Press.
DOI: 10.1093/oso/9780190902704.001.0001

during times of greatest need. This story about the need to call and call upon the *"Mutha'"* has roots in ancestral myths and reaches now into future visions beyond the Anthropocene. It appears in versions too numerous to fully recount, so here I consider only the few that I find particularly meaningful to what's going on in the Anthropocene.

FACING THE "MUTHA'"

In 1981, the National Coalition against Censorship reported that because of its "corrupting influence," the June Jordan poem "Getting Down to Get Over" had been "effaced" from a high-school textbook in Virginia.[3] The stated reason probably was the poem's inclusion of the word "MOTHAFUCKA" in screaming capital letters. But the censors' fear more likely stemmed from the poem's capacity to corrupt faith in *The Man*. Jordan not only loudly calls out the "first primordial/the paradig/dig-matic/dogmatistic mothafucka,"[4] but announces the end of that Mother-effacing paradigm, calling openly upon "MOMMA," the cunctipotent force-source.

"Getting Down to Get Over" is a long poem that touches many interconnected realities. Jordan recognizes Black mothers, people, and families surviving while being cross-cut by racism and sexism within a thieving and pestilent system, marked with lynching, violent policing, and mass incarceration. She invokes the mother, lamenting the lies being told about her: *"Black* Momma/*Black* bitch/*Black* pussy . . . Black Matriarchal Matriarchy/ . . . Hallelujah Saintly/ . . . martyr masochist/(A BIG WHITE LIE)/Momma Momma."[5] Jordan condemns White America's sacrifice of Black women and accompanying devastation of the "house" that is the Earth. Jordan calls out to the "MOMMA," asking for her "help" in together turning away the trouble, turning it around, and making it, this time, turn out differently:

MOMMA MOMMA
momma momma
family face
face of the family alive
. . .
the house on fire/
poison waters/
earthquake/
and the air a nightmare/
turn
turn
turn around the
national gross product
growin
really gross/turn
turn
turn the pestilence away
. . .
hey
turn
my mother
turn
the face of history
to your own
and please be smilin
if you can
be smilin
at the family

momma momma
. . .
teach me to survive my
momma
teach me how to hold a new life
momma
help me

turn the face of history
to your face.[6]

"Getting Down to Get Over" is all at once diagnosis, lamentation, accusation, affirmation, rite of renewal, and invocation of infinite being. It is prayer and prophecy, as Jordan cries out: "let the funky forecast/be the last/one we will ever/want to listen to."[7] In an African-based cosmologic, the state of carnal, fragrant, fructuous *funkiness* is "spiritual oneness with the cosmos," the knowledge "that people are created in harmony with the rhythms of nature and that free expression is tantamount to spiritual and mental health."[8] If the "funky forecast" is the "last one," the weather might just change for the better; the rainbow bridge just might appear after the storm.

Jordan petitions the "momma" to help her to turn the "face of history" to the mother's face. This *history* is not that of *Man* or even humanity, but of the earthbound, a history that Rachel Carson explains, *is* biology:

> I like to define biology as the history of the earth and all its life— past, present, and future. To understand biology is to understand that all life is linked to the earth from which it came; it is to understand that the stream of life, flowing out of the dim past into the uncertain future, is in reality a unified force, though composed of an infinite number and variety of separate lives. The essence of life is lived in freedom. Any concept of biology is not only sterile and profitless, it is distorted and untrue, if it puts its primary focus on unnatural conditions rather than on those vast forces not of man's making that shape and channel the nature and direction of life.[9]

Knowing this truth is to see that the "Mutha'" is the vast force, the source and shaper of knowledge. Some who never forgot that truth, and some who have newly faced it, are working so that "MOMMA" might again be "smilin at the family." Right now, she does not seem to be so inclined.

FACING CHANGE

In 1960, in the first spurt of the Great Acceleration, John Thomas Biggers painted the Earth Mother in his remarkable mural *Web of Life* (Figure 8.1).[10] Biggers, Distinguished Professor of Art at Texas Southern University, was an artist from the Jim Crow South. Like bell hooks, Biggers roots his ecological knowledge in childhood absorption of Black oral traditions and practical experiences. He learned the "interdependence of life" from photosynthesis and explains that he knew, "as early as my high school days why man couldn't continue haphazardly and ruthlessly to exploit, to destroy nature without harming himself. But . . . we didn't refer to this in those days as ecology. My memories of simple folkways and my love of the land, of animal and plant life, as a country boy, gave me inspiration for this mural."[11]

His boyhood memories were later enriched when he received a grant to travel to Africa. Biggers's desire to go to Africa was "revolutionary," scholar and curator Alvia J. Wardlaw writes, in 1950s America, when the nation still was justifying its history of slavery by projecting Africa as the "Dark Continent."[12] Biggers set out seeking "*maamé*, a word in Akan . . . for the Great Mother, the maternal spirit in all things."[13] In Africa, Biggers encountered *maamé* "everywhere, in the shrines to her throughout the countryside, in the women themselves, and in the music of Africa."[14]

FIGURE 8.1 John Thomas Biggers, *Web of Life* (1960). © 2019 John T. Biggers Estate / Licensed by VAGA at Artists Rights Society (ARS), NY. Estate Represented by Michael Rosenfeld Gallery. Casein with egg emulsion on canvas, 72" x 312". Original location: Nabrit Science Hall, Texas Southern University. Present location: Texas Southern University Art Gallery.

He gravitated particularly to the "Fanti and Ashanti" peoples in Ghana and their belief "in male and female entities of God."[15]

Returning to the United States, Biggers took up the great theme of creation but, as he put it, "from a matriarchal point of view; whereas European artists had been motivated to paint creation from a patriarchal point of view."[16] Biggers further noted that the experiences of Africa had "aroused in me deep emotion and thought," leaving him with the goal of producing "a convincing image of *maame*—the fountain of life."[17]

His mural *Web of Life* manifests that image. The colors are mostly olive greens transitioning to browns in a panorama that incorporates the seasons, substrata, ground, sky, and life-sustaining patterns of nature. A great tree is at its core. Under the surface is a naked, fleshy, not markedly feminine Earth Mother, who sits nursing a child in a womb-like enclosure made by the tree's roots. Around her, as if emanating from her, is a spider web of golden sun rays, the sign of the African (male) spider Ananse, "a heroic character in African folklore [who] . . . outwits all."[18] The great tree is filled with multiple species in all states of coming and going, growth and decay. On the ground, just atop the roots, are two humans, female and male. Biggers thus nods to the patriarchal version of creation, but shows the actual "fountain of life," the ambisexual, shining-dark Earth Mother who sits beneath (Figure 8.2).

In a graphite pencil study for the mural, from 1958, Biggers first depicts the Earth Mother nestling her infant, looking down at the child in the conventional manner of Madonna paintings. In a second study, she lifts her head slightly.[19] In a subsequent painting, and then in the final work, the Earth Mother faces us. (Figure 8.2) One side of her visage remains in shadow, the other is alight. She wears an admonitory mien—fierce, formidable, grim, and arresting. Such an "outraged mother," literary scholar Joanne Braxton writes, is a recurring presence in African American creative traditions—an "ancestral figure," one who "feels very keenly every wrong done her children, even to the furthest generations."[20]

FIGURE 8.2 Detail from John Thomas Biggers, *Web of Life.* © 2019 John
T. Biggers Estate / Licensed by VAGA at Artists Rights Society (ARS), NY,
Estate Represented by Michael Rosenfeld Gallery.

Biggers's unsmiling Earth Mother bespeaks a profound distur-
bance not only in Anthropocenic human society, but at the very
center of the "web of life."

   Art historian Olive Theisen writes that Biggers, with his focus
"on the connectedness of human beings with the Earth . . . fore-
shadowed the environmental movement by nearly a decade."[21]

She is right. But with his centering of the scowling, ambisexual, outraged, shining yet dark Mother Earth, as well as his recognition that "creation does not spring down from the sky, rather, it rises up and returns to the earth,"[22] Biggers foreshadows as well the environmental justice movement, ecofeminism, ecowomanism, the revolution against heteropatriarchal sexual and gender binaries, and Afrofuturism.

Afrofuturism's reach embraces theory, performance, historical and speculative fictions, music, and art in an "intersection of imagination, technology, the future, and liberation."[23] In keeping with Africana traditions, Afrofuturism refuses oppositional and hierarchical dualism, centering *maamé*, as explained by Ytasha Womack:

> Afrofuturism is a home for the divine feminine principle, a Mother Earth ideal that values nature, creativity, receptivity, mysticism, intuition, and healing as partners to technology, science and achievement . . . There's a wide-spread belief that humankind has lost a connection to nature, to the stars, to a cosmic sense of self, and that reclaiming the virtues of the divine feminine will lead to wholeness. Many men in the genre embrace the principle as much as the women do.[24]

A complex Afrofuturist invocation of this divine feminine appears in Kevin Sampson's 2015 divinatory mural *Fruit of the Poisonous Tree*, commissioned for the unique exhibit *Anthems for the Mother Earth Goddess*. The Andrew Edlin Gallery in New York City invited seven artists to make works related to ecology and install them right onto the walls, ceilings and floors. Each artist had seven days and nights to complete their piece. Then, when the show closed, the building along with all the works would be demolished.

Sampson is a former New Jersey policeman who says that he consciously carries on the legacy of his civil-rights-activist father. A *New York Times* reviewer exclaims that Sampson's "visionary

'Fruit of the Poisonous Tree' protests almost everything about modern society."[25] Sampson would agree, as he aims "to show the interconnected relations of all of these struggles . . . You can't be for African American rights if you are not for gay rights, woman's rights, the planet's rights . . . immigration rights and workers free from corporate exploitation."[26]

Fruit of the Poisonous Tree (Figure 8.3) is set off with painted curtains (with stitched-on confederate flag symbols), drawn to the side and opening into a cosmic scene of life on Earth, again centered on a great tree, both poisoned and poisonous. Pipes penetrate the bark, draining the life stream. A protruding spigot pours out a portrait of Michael Brown, an unarmed Black youth shot dead by a White policeman in Ferguson Missouri in 2014. Place names are cut into the bark—Staten Island, Newark, Baltimore—referring to locations notorious for such killings of Black men (women, girls, and femmes' names don't appear here, though there are many).[27] At the root of the tree appears the word "Miasma"; these roots turn into hooded Ku Klux Klan figures, bearing names of

FIGURE 8.3 Kevin Sampson, *Fruit of the Poisonous Tree* (2015). Mural by the Rusty Thorn Collective, Kevin Blythe Sampson, Cesar Melgar, Jerry Gant, Manuel Acevedo, James Wilson.

"Monsanto," "RNC," "Keystone," and "Fracking." To the right of the tree is a black rhinoceros, a critically endangered species, with human feet, suggesting the parallel endangerment of African rhino beings and Black American human beings.

On the left is the Earth Mother Goddess—bluish green, partly artefactual, breasted but genitally ambisexual (Figure 8.4), wearing a horned bovine mask reminiscent of Picasso's *Woman in Green*.[28] The mother's arms and torso are sliced into sections and severed from the shoulders; the hands are open, palms facing out. Instead of legs ending in feet, this being bottoms into a sort of spike. Sampson's Earth Mother Goddess pauses before a bluish-green womb-like sphere. Emptied pill capsules intrude on the sphere, which is penetrated in several places by the draining pipes. Still, there are wave motions on its surface. This is the cosmic matrix or what Little Bear calls the "flux." There still is a possibility of rebirth because the sphere contains swimming sperm and those swirling motions indicate potential. A side-wise and red-rimmed mandorla[29]—the vulviform shape that recurs in sacred iconography—appears behind this Earth "Mutha'." Placed on its side, the mandorla becomes a mouth. "Hands up"—the Black Lives Matter slogan against police violence—is inscribed upon those blood-red rimmed lips.

This arresting Mother Earth Goddess is masked and mysterious. Sampson explains, "My mother is Woman and Man, and not quite, she is more than human, less than human, a reaction to her-his Surroundings . . . seeking to create a bubble to shield her-himself and the planet." This Earth Goddess, he continues, is "adjusting to the depletion of the earth by both environmental and social destruction. She is morphing into something new as a way of protecting what is good, so she has to be both caring and brutal in order to survive. She is being fertilized by her desire to endure."[30] *Fruit of the Poisonous Tree* represents the interconnections of social oppressions with consanguineous atrocities committed against peoples, other animals, land, waters, and air. Hate, greed, violence, and power-madness drain and

FIGURE 8.4 Earth Mother Goddess, detail from Kevin Sampson, *Fruit of the Poisonous Tree*.

poison biology-history, inducing a change in the patterns that guarantee continuous rebirth. This is clearly signaled as this self-fertilizing Earth Mother Goddess "morphs," while refusing to show their face.

These inter-resonant works of Jordan, Biggers, and Sampson call to mind writer and Afrofuturist Octavia Butler and her prescription for human behavior in the face of incommensurable Earth changes and a coming collapse of civilization. Butler's paired novels *Parable of the Sower* (1993) and *Parable of the Talents* (1998) are eerily prescient. Butler envisions America in complete spiritual-material chaos as soon as 2024. The ecology is permanently altered by climate change, as horrific fires sweep through California. Society is rife with rape, racism, corrupt police, mass incarceration, pharmaceutically enabled drug abuse, neo-slavery, children who were seized from their parents, increasing corporate control, walled-off communities of the wealthy, caravans of displaced people heading north to flee violence and poverty, and the installation of an authoritarian Christian right-wing government headed by a demagogue proclaiming "Help us to make America great again."[31] In response, the books' formidable protagonist—whom we first meet as a teenage girl—founds a new religion and alternative community, Earthseed.

Lauren Oya Olamina is not conventional. She has heterosexual relationships, but is not traditionally feminine; sometimes she passes as a man, attracting and being attracted to women. Her founding theology and practical advice, collected in *Earthseed: The Books of the Living*, rejects a "super-person"–type deity or "big-daddy-God or a big-cop-God or a big-king-God." Just as firmly, she refuses any deification of nature that understands "nature" to mean whatever people "happen not to understand or feel in control of."[32] Belief in any deity, Butler wisely warns, will "not save you."[33] Only grounded action—political, spiritual, and practical—will. Such action, Butler theorizes, seeks to shape (not to control or overwhelm) the great force of nature, "*Change*," understanding that this also inexorably changes you.

All that you touch
You Change.

All that you Change
Changes you.

The only lasting truth
Is Change.

God
Is Change.[34]

People cannot control change, Butler acknowledges, but we are not powerless. We can, "with forethought and work/ . . . shape God," to some extent, though in the end all must "yield" to the inexorability of the force of "Change."[35]

Although Earthseed does not have conventional goddesses or gods, one notable Yoruban deity is invoked by Olamina's middle name *Oya*—"the name of a Nigerian Orisha—goddess—of the Yoruba people. In fact, the original Oya was the goddess of the Niger River, a dynamic, dangerous entity. She was also goddess of the wind, fire, and death, more bringers of great change."[36] Oya, moreover, according to scholar Judith Gleason, did not stay in her home continent, but traveled to the Americas along "with other African gods, in the heads of worshipers chained in the holds of slave ships."[37]

Oya is known for pointed speech and fiery truth-telling. A traditional praise "sung by a small group of old women, clothed for the occasion in red and purple," goes, in part, "Mother, Oya/ She's the one who employs truth against wickedness/ She stands at the frontier/between life and death." The song hails Oya as the "insatiable vagina" and promises that "Fighting Oya will come into her own."[38] It seems the truth-telling Oya *is* the source-force of the inspired tongue named by M. NourbeSe Philip as a "language of jamettes," one of outspeaking cuntspeakers. Oya is a warrior, fierce and, like Venus-Aphrodite, sometimes bearded. Multiple practitioners and scholars affirm Oya as matron/patron of gender-variant and same sex erotically inclined persons.[39] At her most awesome, Oya is a weather deity, capable of great destruction.

To survive the coming weather events, including climate change, extra-keen perceivers gather knowledge from the ground up and become as adept as possible at influencing change, hoping that the formidable "Mutha'" is not utterly inexorable. Gleason tells us that "the diviners who recite" the praises to Oya "are concerned not only with getting people who ought to be worshipping her to do so, but with keeping Oya herself in line as well."[40] This is precisely the *shaping* of "Change" that Butler describes, as well as a prescription through both ritual and practical activity to renew the patterns that sustain us—lest they and we disappear, as Little Bear prognoses, into "the flux."

In Taoism, too, *change* is the universal principle: When something goes beyond its limits, it inevitably morphs into its opposite.[41] The story of the Earth as "Mutha'" declares that *The Man's* order based in motherfucking has reached its own final frontier and is collapsing. The outcome is unpredictable, but maybe the script can flip and land somewhere better if humans act with forethought and in the name of the "Mutha'" to shape the new birth now in process.

The internationally celebrated artist Wangechi Mutu's animated video *The End of Eating Everything*[42] tells a related story featuring a figure suggestive of a Mother Nature-Earth (played by the musical artist Santigold). The video opens with a glimpse of nebulous space, against which the title appears. A flurry of black birds enter from the bottom left into this grey sky and crash into the title, which falls to sounds of breaking glass. On the right, the profiled face of a woman with long Medusa-like coils of hair slowly enters the frame. Words at the bottom of the frame read: "I never meant to leave/I needed to escape/And now it's so far/Who knows where?/It's been like this for a very long time . . . /It follows me, and I them/We've always been/Hungry, alone and together." The being roars and begins sniffing and eating birds, who pop in bursts of blood. (Figure 8.5) The camera pulls back to reveal her violated and distorted body (Figure 8.6). If she once was spherical, she is no longer. This being is disturbingly shaped and marked with grotesque colorful tumors and eruptions. Wheeling broken mechanisms lie on her surface. Dark feminine human arms with waving hands jut out in multiple places, suggesting bodies lying unquietly beneath. Grey gases spew from several craters, including, we soon see, her anus, finally forming a fog that spirals full screen in an undoing, counter-clockwise direction. As that swirl dissipates, blue skies and white cumulus clouds emerge. From each appears a version of the original being's head, all looking around. (Figure 8.7) Their mouths move in speech, though the words are neither recorded nor transcribed. Mutu's fed-up mother—starved, poisoned and trashed—consumes the consumers and then bursts, ending one world, but now excitedly speaking, possibly bringing new worlds into being.

In an interview along with Mutu, Santigold explains that our consumer culture is "so far on the wrong side" of reasonable behavior that "there is no more room, there's going to be an explosive renewal." Mutu thanks her for so beautifully summing up the intent of the piece.[43] *The End of Eating Everything* mocks the fiction that makes *Man* the only agent and disregards the "Mutha'" as the one who makes and unmakes, comes and goes, changes of her own accord, self-generates and continues to create within the larger context.

NEW WAYS AND WORLDS

Visionary narratives from philosophy through popular culture—for example, Mary Daly's "Lost and Found Continent" and *Black Panther*'s hidden world of *Wakanda*[44]—imagine another place beyond the rapist and consumptive reach of Anthropocene *Man*. This place exists on the other side of a rainbow bridge. It's

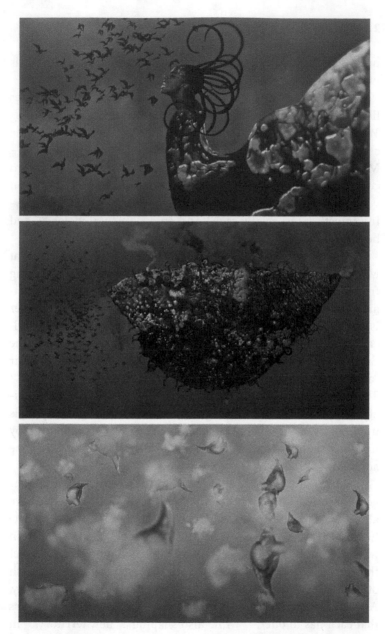

FIGURE 8.5 Stills from Wangechi Mutu. *The End of Eating Everything*, 2013. Animated color video with sound. 8 minutes, 10-second loop. Copyright Wangechi Mutu. Courtesy of the artist, Gladstone Gallery, New York and Brussels, and Victoria Miro.

underneath the one we inhabit or it is right here, but hidden. This place is both internal and external, imaginary and real.

Creative thinker, flautist, and composer Nicole Mitchell takes listeners to such a place in *Mandorla Awakening II: Emerging Worlds*, performed with her Black Earth Ensemble—a name chosen "to honor the feminine source that our lives depend on— Mother Earth."[45] *Mandorla Awakening* features a spoken word component, "Staircase Struggle," written by Mitchell and given voice by artist avery r young, invoking a "dark mother," decrying injustices, especially racist and environmental violence, and beseeching humanity to "find another way."[46]

The album is based upon a short story (unpublished) that Mitchell wrote, "originally called *Mamapolis*," and evolving from "a fascination I had with trying to understand how patriarchy began." As the story developed, she focused less on gender issues and looked more particularly at the ways that "technological advancement is not aligned with caring for ourselves and the Earth." Mitchell honors nondualistic multiplicity and rebukes our "addiction to greed, which compromises our love for each other."[47] The composition takes up the question of what "an advanced society that is in harmony with nature [would] be like." Mitchell calls upon all of us "to quest this question into being," including by unifying the realms that have been so disastrously severed. She advises invoking "new worlds and words."[48]

Mandorla Island is a refuge in an otherwise dying and dangerous world, a place with immanent healing powers that have allowed the culture "to advance without compromising the Earth."[49] Mitchell makes it clear that the belief that "Western ways are 'more advanced' is not really true, and we need to let that go." Her desire is that "we learn to see ourselves in each other," to heal the cuts made between nature and technology, African and Western "to find the common place of goodness that each has to offer,"[50] while also shaping a "new territory"[51] in which to dwell. Previously, Mitchell composed suites honoring Octavia Butler, and, like Butler; she tempers her pessimism with practical understanding of Oyan spirituality and praxis: "I'm African-American, and I have a lot of anxiety living in this world today . . . 'Mandorla' isn't an altogether 'hopeful' piece, though optimism peeks through at times. I tend to be pretty optimistic, because I believe humanity has the power to co-create its future."[52]

Mandorla is an Italian word that means *almond*, and it holds symbolic meanings of hidden truth, the union of opposites, and, most basically, of "the vulva, the yoni, which the Upanishads tells us is 'the symbol of the cosmic waters and the stormy whirlpool of the infinite possibilities of existence' . . . the primordial womb."[53] The upright mandorla appears in the radiant vulviform aura around the Virgen de Guadalupe as well as Jesus and Catholic saints. Lying on its side (as the symbol does in Sampson's mural) the mandorla becomes the symbol of Christ "hidden in the womb."[54] Here, too, is a realization of the feminine divine as encompassing all, including what is understood now as the masculine. Biggers painted the Earth Mother—"*maame*—the fountain of life"—in fusion with Ananse, the Akan male-identified trickster-spider. As someone with a Catholic background, this same entity strikes me as a radiant Black Madonna and a fruitful Green Christ, with the Madonna generating and regenerating the growing, dying, and transforming Christ—an elemental pairing I imagine recurs in any number of traditions, old and new.

Mitchell cites the influence of Riane Eisler's *The Chalice and the Blade*,[55] which traces a historic disastrous shift from egalitarian "chalice" cultures to patriarchal "blade" ones. The mandorla is identified with the chalice, as the symbol appears at the center of two overlapping chalices (as seen on the Chalice Well at the foot of Glastonbury Tor in southern England, a site venerated by modern pagans). The same mandorla sign appears, jarringly, in the MasterCard logo. Graphic designer Maggie Macnab traces the sign to its origin in a "mother circle" and then decodes the credit card logo as signifying the coming together of the opposites of "East and West," now able "to find resolution through the spending power of MasterCard."[56] I see it differently and not so benignly, for the logo is actually proclaiming East meeting West over the "master's" now global capitalization of the womb of the Earth Mother.

Mitchell inspires us to "find another way" outside the master's purview (and the reach of his card). She evokes what it will feel like when we get there and issues this invitation:[57]

> Some had our fortune to stick our hands in the black soil. We instinctively learned that dark matters. That's where the mind is free. Birds sing interlocking songs of imagination. An image nation of endless possibility. There is a place of innovation, of improvisation, of impossible. That's where our survival is. Many have dipped to drink its power. Darkness is the beauty and will always be. New worlds, new worlds and words, new worlds and words can change this illusionary one. Enter.

Mitchell's composition and the Black Earth Ensemble performance of *Mandorla Awakening II: Emerging Worlds* calls explicitly to "dark mother" and calls on her, petitioning infinity, finding another way, venturing to an old-new *terra*-territory—a place of many names, including now *Mandorla Island*. There we learn that much of what has passed as reality is an illusion, that what poses as progress is not, and that much of what we have been conditioned to find bad and ugly—like the dark, the dirt, the animal, the "flawed," and the funky—are actually "the beauty" and also "where our survival" lies. That beauty is not something reserved for a select, privileged, and supposedly perfect few. Rather, it is the birthright and being of all the earthbound, made by the "Mutha'" in her image.

CALLING ON THE "MUTHA'"

Wildly popular animated films like *Frozen*,[58] *Moana*,[59] and *Shrek*[60] all share elements of the Call Your "Mutha'" story, as do revelatory novels by Linda Hogan (*Power)* and Jesmyn Ward (*Salvage the*

Bones).[61] In all these, protagonists in the face of eco-apocalyptic dangers act to call back an estranged life force-source. The same theme appears throughout the ancient oral tradition. Apparently, humankind has been in this sort of danger before, or seers have anticipated it, and these stories instruct on what is best to do.

A Hopi version[62] concerns a magic hummingbird, formed from a sunflower stalk by a boy to distract his little sister from their imminent starvation in a dying world. The girl tosses the sunflower-stalk-turned bird into the air four times, unexpectedly animating the being. The hummingbird leaves the surface world to go down through subsequent worlds to a fourth and flowering one. There the bird being calls upon the mothering power of Muy'ingwa, "the god of fertility and germination," beseeching Muy'ingwa to once again return to the surface and renew the world. Muy'ingwa understands and agrees to again be present in the upper world, undertaking a return journey upward through the worlds, offering prayer—full concentration—at every level. Muy'ingwa's return to the surface results in the needed renewal. The rain falls, the land turns green, and beauty and life are restored.

In a Shinto story, the sun goddess Amaterasu, in response to a rape, retreats to a cave. In her absence, the world grows terminally cold and dark. The elemental beings gather in desperation. Relief comes only when the old crone Uzume performs a risible striptease at the mouth of the cave, exposing her genitals to the merriment of all. Amaterasu, hearing the laughter, comes out and beholds, not only the revealed Uzume, but also her own radiance in a mirror that was set up to catch Amaterasu's eye. Face to face with the source of life, she agrees to come back to the world.[63]

The Greek story of Demeter and Persephone is similar. These two originate as aspects of one Earth deity, with above- and below-world presences. In a later, patriarchal version, Persephone is raped and abducted by Hades. In response, the Green Mother Demeter turns herself into a grim and grey crone, mourning and withdrawing all fertility from the land. Life resurges only after an old and raucous sage calls upon her. Sometimes it is the lame Iambe, who sings amusing dirty songs. Other times, it is the aged and bawdy Baubo (a name meaning "belly" and sometimes "dildo") who dances and lifts her skirt, exposing her genitals. The dirty dance and funky play activate the erotic and re-energize the goddess,[64] who smiles, relents, and returns.

These ancient rituals of renewal—combining age, gender ambiguity, and the erotic—recall those West African rites performed by postmenopausal Mothers who center the vulva as a "living altar," rites that are the last resort for a society in times of greatest social emergency.[65] They also remind me of a story of genital power told by Zora Neale Hurston regarding her time of initiation into Haitian Vodou. Her teacher has asked her, " 'What is the truth'? . . . and knowing that I could not answer him he answered himself through a Voodoo ceremony in which the Mambo, that is the priestess, richly dressed is asked this question ritualistically. She replies by throwing back her veil and revealing her sex organs. The ceremony means that this is the infinite, the ultimate truth. There is no mystery beyond the mysterious source of life."[66]

These words of Hurston, once again invoking "the infinite," swirl me back to the beginning of this book and then back around again to its conclusion. Social-ecological calamity is happening right now to so many and its spread seems inevitable. Seers, though, point to possible survival as well as eventual renewal. The story going round about calling the "Mutha'" instructs that motherfucking must be foresworn; all forms of rape and rapism must end; and there must be express ritual feeding of the green. This includes practical activity, such as establishing environmental and reproductive justice, developing green sciences and technologies, and participating in renewal via "the medium of prayer" (Momaday)—focused, concentrated attention, individual and group. The "Mutha'" as the "earth parent" (Tuhiwai Smith) is neither mysticism nor metaphor, but the basis for developing the scientific and practical knowledge that allows being beyond the Anthropocene. In calling, and calling upon, the "Mutha,'" one becomes a revolutionary mother, shaping the incommensurable changes now happening in ways that *make life* (in LaDuke's sense).

Some years ago, I called upon the mothering power, though I didn't know it at the time. I was at the 2009 Southeast Women's Studies Association conference in Boone, North Carolina. Starhawk was giving a workshop and, for its final turn, she asked participants to meditate and envision a future that they would want to live in and would work to manifest. I usually don't have much luck at meditation, but I tried. Strangely enough, I did go someplace else. I walked across a bridge over a waterway and found a community with small, rounded dwellings, low to the ground, and with surrounding greenery and land. There, approaching to welcome me was a vividly green and very tall being with large, flappy, bat-wing-like, leafy ears—just like the face of the green Dracula in the poster for the 1928 theatrical production (which I had seen previously). This green being was holding out a hand/limb to me in greeting, and was wearing a long brown coat, making them something like a living tree, even a Tree of Life.

The meanings of such experiences are always mysterious and unfolding. Still, here are some impressions. Calling and calling upon the "Mutha'" opens the possibility of again being able to approach, greet, meet, and even touch the Tree of Life. Together, can we call, and call upon, the "Mutha'"? Can we (materially and/or imaginatively, as best suits) carry a seed, *be* a seed, cross a bridge, and take the outstretched hand? Can we smile back at that cosmic face, listen to the low sounds, and touch the ground? And then, sticking "our hands in the black soil" (Mitchell), plant that?

CONVOCATION

"The action of calling together," from Latin convocāre to call together; < con- together + vocāre to call." (OED)

Call Your "Mutha'". Jane Caputi, Oxford University Press (2020). © Oxford University Press.
DOI: 10.1093/oso/9780190902704.001.0001

CODA

"Gather and Vote"

IN ONE OF THE DREAMS that guided me to this book, I encountered a rough, black stone about a foot high, standing on the ground. In a flash, the stone animated, and (reflecting my Irish heritage) there now stood a green-garbed and red-capped leprechaun. After we exchanged polite introductions, I asked him if I would be permitted a question. When he nodded yes, I asked what could done in these dire times. He responded that "scattering is a problem" and succinctly advised, "Gather and vote." With that, he disappeared down a hole that had opened in the ground.[1]

These parting words were at first a puzzlement. Gathering is the basis of social movements and also the principle underlying the oneness of being, so that made sense, but I wasn't sure about voting since I associated this only with the exercise of civic rights and responsibility at the polls. However important, that couldn't be the only message. So I went to the dictionary and looked up *vote* and found that its first, if now rare, meanings were "to vow (to do something)" and "to devote religiously" (*OED*). *Vote*, I also learned, derives from *vovēre*, "to vow, to desire," the same etymological root as *devotion*. To be devoted is to give "reverence" as well as a "gift" and to be deeply "attached" (*OED*) to a person or a cause. To *gather and vote* in these dire times of the Anthropocene, then, is to get it together (individually and with one's community) to do something spiritually and politically,

Call Your "Mutha'". Jane Caputi, Oxford University Press (2020). © Oxford University Press.
DOI: 10.1093/oso/9780190902704.001.0001

that is, to be devoted—vowing love, allegiance, and attachment, including by returning gifts, to the person and cause who is Mother Nature-Earth, our planet/ourselves.

The "psychopaths" (Bambara) running and ruining the planet deny attachment to Mother Nature-Earth. They aim to separate everyone from land and each other. This forced separation blocks cunctipotence, allowing oppressors to split, kill, destroy, and burn without compassion, fomenting stress, alienation, illness, trauma, hurt, grief, and premature death for their targets. At the same time, it encourages apathy, belief in one's own powerlessness, as well as denial and complicity. All of us, if we have not been doing so, need to face the utter-terrestrial reality that humans *be* only because we are in ongoing relation to the mothering power. This is perhaps the most needful of knowledges in the Anthropocene. This knowing cannot be abstract but must be felt with the heart and put into practice.

Life *is* a gathering, a convocation, a ceremony of continually making life in concert with all being. Consider it, then, our most profound civic birthright and responsibility to recognize the larger Earth-community of which we are a part, to share the common goods fairly, to treat all beings with respect, and to gather and vote for Nature-Earth, to gather and vote for the "Mutha'."[2]

POETRY CREDITS

ACKNOWLEDGMENTS

My thanks go first of all to Ruth O'Brien for inviting me to participate in this series and for being my inspiration and advocate along the way. I am most grateful to Marilou Awiakta, who wrote the poem that provided the inspiration and now the epigraph for this book and I have been enriched and refreshed by our conversations over many years. Suzanne Kelly, with immeasurable generosity and acumen, read the manuscript in early and late versions and gave me vital help, attention, encouragement, and advice. Toni Francis did the same and I benefited, again immeasurably, from her suggestions and guidance. My colleague Nicole Morse read a late draft and was another excellent reader, giving me helpful feedback. Oxford University Press editor Angela Chnapko was gracious and most astute. I am indebted to the anonymous and exceptionally helpful readers she chose for my proposal, particularly the extensive one provided by the reader of the full draft. Mary Buchanan provided expert line-editing, working with me as I polished my final draft.

Friends and colleagues offered support, read a section, and/ or suggested materials (including ads) along the way and I thank Rain, Sika Dagbovie-Mullins, Simone Clunie, Jeanette Coleman, Janell Hobson, Kristin Lieb, Helene Vann, Annie Eysturoy, Carol

Prusa, Josephine Beoku-Betts, Karen Leader, Wanda Teays, Phoebe Godfrey, Susan Love Brown, Kathy Merlock Jackson, Sofia Honekman, Selena Quiros, Annette Evans, Dani Orias, Ana Rodriguez, and Peter Cava. Egla Martínez Salazar generously instructed me on the life and work of Berta Cáceres.

In 2017–2018, I spent an academic year in residence as an interdisciplinary scholar at Merrimack College in Massachusetts, while writing. I thank Debra Michals, the Director of Women and Gender Studies, for her collegiality and support and am most grateful to Beth Walsh who made that visit possible by inviting me to share her home. My program at Florida Atlantic University, Women, Gender and Sexuality Studies, has offered me unremitting support, and I especially thank our director at the time, Barclay Barrios, for facilitating that and the Deans of my college during the period, first Heather Coltman and then Michael Horswell. As always, I am grateful to my family and friends, including my partner Roger Messenger, my parents and my siblings, nieces, and nephew for their humor, love, and encouragement.

I live, work, and wrote most of this book on the lands of the Tequesta and Seminole nations. I thank them and I thank the spirit of the land for sustenance and inspiration. I also thank those forces helping and driving me along the circuitous, dark, unmarked and windy ways, the Black Madonna and the Green Guide.

NOTES

Introduction

1. Geneva Smitherman, *Black Talk: Words and Phrases from the Hood to the Amen Corner*, Revised Edition (New York: Houghton Mifflin, 2000), 133.

2. Gabrielle Hecht, "The African Anthropocene," *Aeon*, Nov. 2013, https://aeon.co/users/gabrielle-hecht; Donna J. Haraway, *Staying with the Trouble: Making Kin in the Chthulucene* (Durham, NC: Duke University Press, 2016), 40; Anna Tsing, Heather Swanson, Elain Gan, and Nils Bubandt, eds., *Arts of Living on a Damaged Planet* (Minneapolis: University of Minnesota Press), 2017; Raj Patel and Jason W. Moore, *A History of the World in Seven Cheap Things* (Oakland: University of California Press, 2017); Laura Pullido, "Racism and the Anthropocene," in *Future Remains: A Cabinet of Curiosities for the Anthropocene*, ed. Gregg Mitman, Marco Armiero, and Robert S. Emmett (Chicago: University of Chicago Press, 2018), Kindle ed.; Joanna Zylinska, *The End of Man: A Feminist Counterapocalypse* (Minneapolis: University of Minnesota Press, 2018).

3. Mary Daly first used the word *rapism* in *Beyond God the Father: Toward a Philosophy of Women's Liberation* (Boston: Beach Press, 1973), 194. Subsequently, she and I defined

rapism as "the fundamental ideology and practice of patriarchy, characterized by invasion, violation, degradation, objectification, and destruction of women and nature." Mary Daly, with Jane Caputi, *Websters' First New Intergalactic Wickedary of the English Language* (Boston: Beacon Press, 1987), 91.

4. *Heteropatriarchy* is "an umbrella term to mean the intertwined systems of patriarchy and heterosexism to include its manifestations as heteronormativity, transphobia, and cisnormativity." Leanne Betasamosake Simpson, *As We Have Always Done: Indigenous Freedom through Radical Resistance* (Minneapolis: University of Minnesota Press, 2017), 253n.

5. Susan Griffin, *Pornography and Silence: Culture's Revenge against Nature* (New York: Harper Colophon, 1981).

6. *Oxford English Dictionary* (New York: Oxford University Press, 1989–). All subsequent references will be noted as (*OED*) in the text.

7. Jesse Sheidlower, ed., *The F-Word*, 2nd ed. (New York: Random House, 1999), 117, 124.

8. "Motherfucker," *Urban Dictionary*, https://www.urbandictionary.com/define.php?term=mother%20fucker.

9. Lean Lakshmi Piepzna-Samarasinha defines *femme* in part as "one of a million kinds of queer femme or feminine genders . . . [in] a revolutionary gender universe." See "A Modest Proposal for a Fair Trade Emotional Labor Economy," *Bitch*, no. 75 (Summer 2017): 21–26, esp. 21.

Chapter 1

1. Steve Connor, "The State of the World? Is It on the Brink of Disaster, *The Independent* (UK), March 30, 2005. Millennium Ecosystem Assessment (Program), *Ecosystems and Human Well-Being* (Washington, DC: Island Press, 2005).

2. Marvin Gaye, "What's Going On," recorded in 1971, released on the studio album *"What's Going On"*, Tamla Records, May 21, 1971.

3. Jane Caputi, "Mother Earth Meets the Anthropocene: An Intersectional Ecofeminist Analysis," *Systemic Crises: Race,*

Class, Gender and Global Climate Change, ed. by Phoebe C. Godfrey and Denise Torres (New York: Routledge, 2016), 20–33.

4. Catherine M. Roach, "Mother Nature Imagery," in *Encyclopedia of Religion and Nature*, ed. Bron R. Taylor (New York: Thoemmes Continuum, 2005), 1107–1110, esp. 1107.

5. Tamra Andrews, *A Dictionary of Nature Myths* (New York: Oxford, 1998), 61.

6. Francis, *Laudato Si'* (Washington, DC: United States Conference of Catholic Bishops, 2015), 3.

7. Elisabetta Povoledo and Gaia Pianigiani, "Vatican Clarifies the Rules for Cremation," *New York Times*, Oct. 25, 2016.

8. Carolyn Merchant, *The Death of Nature: Women, Ecology and the Scientific Revolution* (San Francisco: Harper and Row, 1980), 41.

9. Micki McElya, *Clinging to Mammy: The Faithful Slave in Twentieth-Century America* (Cambridge, MA: Harvard University Press, 2007).

10. *Feed the Green: Feminist Voices for the Earth*, a film by Jane Caputi, produced by Susan Rosenkranz (New York: Women Make Movies, 2015), DVD, http://www.wmm.com/filmcatalog/pages/c934.shtml.

11. Universal Declaration on the Rights of Mother Earth and Climate Change (UDRME), 2010. World People's Conference on Climate Change and the Rights of Mother Earth, April 22, 2010, https://pwccc.wordpress.com/programa/.

12. Elizabeth Kolbert, "Enter the Anthropocene: Age of Man," *National Geographic*, March 2011, 60–85.

13. Christian Schwägerl, *The Anthropocene: The Human Era and How It Shapes Our Planet*, trans. Lucy Renner Jones (Santa Fe, NM: Synergetic Press, 2014), 10.

14. Paul J. Crutzen and Christian Schwägerl, "Living in the Anthropocene: Toward a New Global Ethos," *Yale Environment 360*, Jan. 24, 2011, http://e360.yale.edu/feature/living_in_the_anthropocene_toward_a_new_global_ethos/2363/.

15. Koert van Mensvoort, "Anthropocene for Dummies," Feb. 8, 2011, *Next Nature Network* (blog), https://www.nextnature.net/2011/08/anthropocene-for-dummies/.

16. George Church and Ed Regis, *Regenesis: How Synthetic Biology Will Reinvent Nature and Ourselves* (New York: Basic Books, 2014), 143.

17. Mark Lynas, *The God Species: Saving the Planet in the Age of Humans* (Washington, DC: National Geographic, 2011).

18. Will Steffen, Paul J. Crutzen, and John McNeill, "The Anthropocene: Are Humans Now Overwhelming the Great Forces of Nature?," *Ambia* 38, no. 8 (2007): 614–621.

19. Sylvia Wynter, "Unsettling the Coloniality of Being/Truth/Power/Freedom: Towards the Human, after Man, Its Overrepresentation—an Argument," *CR: The New Centennial Review* 3, no. 3 (2003): 257–337, esp. 260.

20. Clarence Major, *Juba to Jive: A Dictionary of African-American Slang* (New York: New World Paperbacks, 1970), 296.

21. *Missy* is the name of the abusive White woman in *Get Out*, written and directed by Jordan Peele (Universal Pictures, 2017), film.

22. Lisa Kemmerer, introduction to *Sister Species: Women, Animals, and Social Justice*, ed. Lisa Kemmerer (Urbana: University of Illinois Press), 1–43, esp. 17.

23. Maria Mies, *Patriarchy & Accumulation on a World Scale: Women in the International Division of Labour* (New York: Zed Books, 1998), 75.

24. Shamara Shantu Riley, "Ecology Is a Sistah's Issue Too," in *Ecofeminism and the Sacred*, ed. Carol J. Adams (New York: Continuum, 1993), 191–204, esp. 192.

25. Hilary Malatino, *Queer Embodiment: Monstrosity, Medical Violence, and Intersex Experience* (Lincoln: University of Nebraska Press, 2019).

26. Simpson explains that *2SQ* "is an umbrella term . . . to refer to all Indigenous Two Spirit, lesbian, gay, bisexual, pansexual, transgender, transsexual, queer, questioning, intersex, asexual, and gender-non-conforming people." Simpson, *As We Have Always Done*, 255.

27. Toni Cade Bambara, "What It Is I Think I'm Doing Anyhow," in *Calling Home: Working-Class Women's Writings: An Anthology*, ed. Janet Zandy (New Brunswick, NJ: Rutgers University Press, 1990), 321–325, esp. 321.

28. Ruth Wilson Gilmore describes racism as "the state-sanctioned or extra-legal production and exploitation of group-differentiated vulnerability to premature death." Ruth Wilson Gilmore, *Golden Gulag: Prisons, Surplus, Crisis, and Opposition in Globalizing California* (Berkeley: University of California Press, 2007), 28.

29. *Violence on the Land, Violence on Our Bodies: Building an Indigenous Response to Environmental Violence*, A Partnership of Women's Earth Alliance and Native Youth Sexual Health Network 2016), 63. This study is coauthored by Erin Marie Konsmo (Métis/Cree) from the Native Youth Sexual Health Network and A.M. Kahealani Pacheco (Kanaka Maoli) from the Women's Earth Alliance. http://landbodydefense.org/uploads/files/VLVBReportToolkit2016.pdf.

30. Elizabeth Povinelli, "The Three Figures of Geontology," in *Anthropocene Feminism*, ed. Richard Grusin (Minneapolis: University of Minnesota Press, 2017), 49–64, esp. 54–55.

31. Damian Carrington, "World on Track to Lose Two-Thirds of Wild Animals by 2020," *The Guardian*, Oct. 26, 2016.

32. Damian Carrington, "How the Domestic Chicken Rose to Define the Anthropocene," *The Guardian*, Aug. 21, 2016.

33. Lisa Kemmerer, "Appendix: Factory Farming and Females," in *Sister Species: Women, Animals, and Social Justice*, ed. Lisa Kemmerer (Urbana: University of Illinois Press, 2011), 173–185, esp. 178–181. Patel and Moore, *History of the World*, 3–5.

34. Carrington, "How the Domestic Chicken Rose"; Lisa A. Kemmerer, *Eating Earth: Environmental Ethics and Dietary Choice* (New York: Oxford University Press, 2014).

35. Jeremy Davies, *The Birth of the Anthropocene* (Berkeley: University of California Press, 2016), 7.

36. Tsing et al., *Arts of Living on a Damaged Planet*, G-3.

37. Apple iPhone XS TV commercial, "Shot on iPhone XS: Don't mess with Mother," 2019, viewable at iSpot.tv, https://www.ispot.tv/ad/IQDx/apple-iphone-shot-on-iphone-xs-dont-mess-with-mother.

38. Susan Sontag, *On Photography* (New York: Farrar, Straus and Giroux, 1977), 14.

39. Steffen, Crutzen, and McNeill, "Anthropocene: Are Humans Now Overwhelming?," 614–621.

40. Sigmund Freud developed his idea of penis envy in several works, beginning with "Some Psychical Consequences of the Anatomical Distinction between the Sexes" (1925), *The Standard Edition of the Complete Psychological Works of Sigmund Freud*, vol. 19 (London: Hogarth Press, 1953), 243–258.

41. Karen Horney, *Female Psychology* (New York: W. W. Norton, 1967). See also Mary D. Garrard, *Brunelleschi's Egg: Nature,*

Art, and Gender in Renaissance Italy (Berkeley: University of California Press, 2010), 54–56l; Daly with Caputi, *Websters' First New Intergalactic Wickedary*, 215–216.

42. Chevrolet "Snow" campaign by McCann, aired July 2010, viewable at Ads of the World, https://www.adsoftheworld.com/media/print/chevrolet_snow.

43. For example, see Susan Griffin, *Woman and Nature: The Roaring Inside Her* (New York: Harper and Row, 1980); Delores Williams, "Sin, Nature, and Black Women's Bodies," in *Ecofeminism and the Sacred*, ed. Carol Adams (New York: Continuum, 1993), 24–29; Karen J. Warren, ed., *Ecofeminism: Women, Culture, Nature* (Bloomington: Indiana University Press, 1997).

44. Diana E. H. Russell, *The Politics of Rape: The Victim's Perspective* (New York: Stein and Day, 1975), 267.

45. Christina Sharpe, *In the Wake: On Blackness and Being* (Durham, NC: Duke University Press, 2016), 16.

46. Winona LaDuke, "A Society Based on Conquest Cannot Be Sustained," in *The New Resource Wars: Native and Environmental Struggles against Multinational Corporations*, ed. Al Gedicks (Boston: South End Press, 1993), ix–xv, esp. x.

47. Merchant, *Death of Nature*, 7.

48. Rachel Carson, *Silent Spring* (Boston: Houghton Mifflin, 1973).

49. Rachel Carson, "Of Man and the Stream of Time," in *Rachel Carson: Silent Spring and Other Writings on the Environment*, ed. Sandra Steingraber (New York: Library of America, 2018), 421–427, esp. 421.

50. Carson, *Of Man and the Stream of Time*, 421.

51. Scott A. Bonn, "The Mass Murder Suicide Connection," *Psychology Today*, Feb. 22, 2018, https://www.psychologytoday.com/us/blog/wicked-deeds/201802/the-mass-shooting-suicide-connection.

52. Naomi Klein, *On Fire: The (Burning) Case for a Green New Deal* (New York: Simon and Schuster, 2019), 45.

53. Caputi, *Goddesses and Monsters*.

54. Mark Lynas, *The God Species: Saving the Planet in the Age of Humans* (Washington, DC: National Geographic, 2011).

55. Daly, *Beyond God the Father*, 13.

56. Linda Hogan, *Dwellings: A Spiritual History of the Living World* (New York: Touchstone, 1996), 85–86.

57. This ad campaign originated in 2009. The ad is ©2010 P&G.
58. Jean Chevalier and Alain Gheerbrant, *The Penguin Dictionary of Symbols*, trans. John Buchanan-Brown (New York: Penguin Books, 1996), 453–454: "The hidden qualities of green derive from the fact that it contains red . . . the fertility of all 'work' arises from [these]."
59. Robin Wall Kimmerer, "Returning the Gift," *Minding Nature* 7, no. 2 (Spring 2014), https://www.humansandnature.org/ returning-the-gift.
60. Emma Goldman and Max Baginski, "Mother Earth," *Mother Earth* 1, no. 1 (March 1906): 1–2.
61. Greg Cajete, *Native Science: Natural Laws of Interdependence* (Santa Fe, NM: Clear Light, 2000), 75.
62. Leroy Little Bear, foreword to Cajete, *Native Science*, ix–xii, esp. xi.
63. Cited in Regina Cochrane, "Climate Change, *Buen Vivir*, and the Dialectic of Enlightenment: Toward a Feminist Critical Philosophy of Climate Justice," *Hypatia* 29, no. 3 (2014): 576–598, esp. 587.
64. Olivia Harris, "The Earth and the State: The Sources and Meanings of Money in Northern Potosi Bolivia," in *Money and the Morality of Exchange*, ed. J. Paray and M. Block (Cambridge: Cambridge University Press, 1989), 232–268, esp. 250.
65. Alex Cross, quoted from a personal communication in the tool-kit section of *Violence on the Land, Violence on Our Bodies*, 5.
66. Linda Tuhiwai Smith, *Decolonizing Methodologies: Research and Indigenous Peoples*, 2nd ed. (New York: Zed Books, 2012), 77.
67. Smith, *Decolonizing Methodologies*, 13.
68. Nicholas Casey, "Climate Change Claims a Lake and an Identity," *New York Times*, July 7. 2016.
69. See, for just one example, the T-shirts for sale online at Redbubble, https://www.redbubble.com/shop/love+your+mother+ earth+t-shirts.
70. Lavinia White, "Listening to Our Elders, *Indigenous Woman* 1, no. 2 (1992): 13–14.
71. Roger Dale and Susan Robertson, "An Interview with Boaventura de Sousa Santos," *Globalization, Societies, and Education* 2, no. 2 (2004): 147–160. De Sousa Santos speaks of an "ecology of

knowledges," where "there is no global social justice without global cognitive justice."

72. Toni Morrison, "Rootedness: The Ancestor as Foundation," in *Black Women Writers (1950–1980)*, ed. Mari Evans (New York: Anchor, 1984), 339–345, esp. 342.

73. Zoe Todd, "An Indigenous Feminist's Take on the Ontological Turn: 'Ontology' Is Just Another Word for Colonialism," *Journal of Historical Sociology* 29, no. 1 (2016): 4–22.

74. L. H. Stallings, *Funk the Erotic: Transaesthetics and Black Sexual Cultures* (Urbana: University of Illinois, 2015).

75. Emma Pérez, "Irigaray's Female Symbolic in the Making of Chicana Lesbian Sitios y Lenguas (Sites and Discourses)," in *The Lesbian Postmodern*, ed. Laura Doan (New York: Columbia University Press, 1994), 104–117, esp. 114.

76. Starhawk, *The Spiral Dance: A Rebirth of the Ancient Religion of the Goddess* (San Francisco: HarperOne, 1999). (This is the 20th anniversary edition.).

77. Gloria Anzaldúa, *Interviews/Entrevistas*, ed. AnaLouise Keating (New York: Routledge, 2000), 178.

78. Anthony J. Nocella II, Judy K. C. Bentley, and Janet M. Duncan, "Introduction: The Rise of Eco-Ability," in *Earth, Animal, and Disability Liberation: The Rise of the Eco-Ability Movement*, ed. Anthony J. Nocella II, Judy K. C. Bentley, and Janet M. Duncan (New York: Peter Lang, 2012), xiii–xxii.

79. Alexis Pauline Gumbs, China Martens, and Mai'a Williams, eds. *Revolutionary Mothering: Love on the Front Lines* (Oakland, CA: PM Press, 2016).

80. For critiques of positivism, see Carolyn Merchant, "Science and Worldviews," in *Radical Ecology: The Search for a Livable World* (New York: Routledge, 1992), 41–60; Patricia Hill Collins, *Black Feminist Thought: Knowledge, Consciousness, and the Politics of Empowerment*, 2nd ed. (New York: Routledge, 2000), 255–256, 265–266.

81. Robin Wall Kimmerer, *Gathering Moss: A Natural and Cultural History of Mosses* (Corvallis: Oregon State University Press, 2003), vii.

82. Mary Daly speaks of "Syn-Crone-icities" as "startling bursts of communication" from other than human nature. Mary Daly,

Quintessence . . . Realizing the Archaic Future (Boston: Beacon Press, 1998), 203.

83. Emily Brontë, *Wuthering Heights* (New York: Modern Library, 1950), 84.
84. Cajete, *Native Science*, 71.
85. Simon J. Ortiz, "Indigenous Language Consciousness: Being, Place, and Sovereignty," in *Sovereign Bones: New Native American Writing*, ed. Eric Gansworth (New York: Nation Books, 2007), 135–147.
86. Pádraig Ó Tuama, "Belonging Creates and Undoes Us Both," interview, *On Being*, National Public Radio, March 2, 2017, https://onbeing.org/programs/padraig-o-tuama-belonging-creates-and-undoes-us-both-mar2017/.
87. Vine Deloria, Jr., "Reflection and Revelation: Knowing Land, Places and Ourselves" in *The Power of Place Sacred Ground in Natural and Human Environments: An Anthology*, ed. James A. Swan (Wheaton, Illinois: Quest Books, 1991), 28-40, esp. 29.
88. Deloria, Jr., "Reflection and Revelation," 37.
89. Zora Neale Hurston, "Go Gator and Muddy the Water," in *Go Gator and Muddy the Water: Writings by Zora Neale Hurston from the Federal Writers' Project*, ed. Pamela Bordelon (New York: W. W. Norton, 1999), 68–79, esp. 69–70. Hurston also uses the word "control" to describe people's ability to work with their surroundings, but her context makes it clear that she does not mean by this what *Man* means—enslavement and spirit breaking.
90. "History of White Earth," White Earth Nation, http://www.whiteearth.com/history.html.
91. Cajete, *Native Science*, 75.
92. Jack D. Forbes, "Indigenous Americans: Spirituality and Ecos," *Daedalus* 130, no. 4 (2001): 283–300, esp. 283. http://www.jstor.org/stable/20027728.
93. Jack D. Forbes, "Nature and Culture: Problematic Concepts for Native Americans," in *Indigenous Traditions and Ecology*, ed. John A. Grim (Cambridge, MA: Harvard University Press, 2001), 103–104, esp. 107.
94. Cited in Joseph Stromberg, "What Is the Anthropocene and Are We in It," Smithsonian.com, January 2013, https://www.smithsonianmag.com/science-nature/what-is-the-anthropocene-and-are-we-in-it-164801414/#J1IFxqqSY6TF Wmah.99.

95. John Mohawk, "From Miracle to Magic: Spirituality as Political Consciousness," in *Nature's Operating Instructions: The True Biotechnologies*, ed. Kenny Ausubel with J. P. Harpignies (San Francisco: Sierra Club Books, 2004), 199–205, esp. 204.
96. Mohawk, "From Miracle to Magic," 203.
97. Mohawk, "From Miracle to Magic," 205.
98. Jack D. Forbes, "Nature and Culture: Problematic Concepts for Native Americans," in *Indigenous Traditions and Ecology*, ed. John A. Grim (Cambridge, MA: Harvard University Press, 2001), 103–104, esp. 107.
99. Patrick Hamilton, "Welcome to the Anthropocene," in *Future Earth: Advancing Civic Understanding of the Anthropocene*, ed. Diana Dalbotten, Gillian Roehrig, and Patrick Hamilton (Washington, DC: American Geophysical Union and John Wiley & Sons, 2014), 1–8, esp. 1.
100. Kate Kaye, "Tangled Up in Big Blue: IBM Replaces Smarter Planet with . . . Bob Dylan," *Ad Age*, Oct. 6, 2015.
101. *Webster's Third New International Dictionary of the English Language, Unabridged* (Springfield, MA: Merriam-Webster, 1986). All subsequent references in the text (*Webster's*) refer to this source.
102. Bruno Latour speaks of "the earthbound" in multiple ways in *Facing Gaia: Eight Lectures on the New Climatic Regime*, trans. Catherine Porter (Medford, MA: Polity Press, 2015).
103. bell hooks, "Earthbound: On Solid Ground," in *The Colors of Nature: Culture, Identity, and the Natural World*, ed. Alison H. Deming and Lauret E. Savoy (Minneapolis, MN: Milkweed Editions, 2002), 67–71.
104. Monica Sjöö and Barbara Mor, *The Great Cosmic Mother: Rediscovering the Religion of the Earth*, 2nd ed. (San Francisco: HarperOne, 1991), 410.
105. Smitherman, *Black Talk*, 133.

Chapter 2

1. Alice Walker, "Preface to the Tenth Anniversary Edition." *The Color Purple* (New York: Harcourt, 1992), xi.

2. Alice Walker, "The Only Reason You Want to Go to Heaven," *On the Issues*, Spring 1997, https://www.ontheissuesmagazine.com/1997spring/sp97walker.php.

3. Eric Partridge, *A Dictionary of Slang and Unconventional English* (New York: Routledge, 1984), 278.

4. Emma L. E. Rees, *The Vagina: A Literary and Cultural History* (New York: Bloomsbury, 2013); Inga Muscio, *Cunt: A Declaration of Independence* (Seattle, WA: Seal Press, 2002).

5. Tracy Chapman, "Rape of the World," 1995, Elektra Entertainment for the United States, New Beginning, track 7, 1995, compact disc.

6. Silvia Federici, *Caliban and the Witch: Women, the Body, and Primitive Accumulation* (New York: Autonomedia, 2004), 88–89.

7. Heinrich Kramer and J. Sprenger, *The Malleus Maleficarum*, trans. Montague Summers (New York: Dover, 1971), 47.

8. Mary Daly, *Beyond God the Father*, 194; Andrea Dworkin, *Woman Hating* (New York: E. P. Dutton, 1973), 95.

9. Lynn White, "The Historical Roots of Our Ecological Crisis," in *This Sacred Earth: Religion, Nature, Environment*, 2nd ed., ed. Roger S. Gottlieb (New York: Routledge, 2004), 192–202.

10. Federici, *Caliban and the Witch*; Raj Patel and Jason W. Moore, *A History of the World in Seven Cheap Things* (Oakland: University of California Press, 2017).

11. Theodore Allen, *The Invention of Whiteness*, 2 vols., 2nd ed. (Brooklyn, NY: Verso, 2012). Andrew Krinks, "Collective Ownership of the Earth Forever and Ever," *Daily Theology (DT)* (blog), Oct. 27, 2017, https://dailytheology.org/2017/10/27/collective-ownership-of-the-earth-forever-and-ever-amen/.

12. Patel and Moore, *History of the World*.

13. Mary Daly, *Pure Lust: Elemental Feminist Philosophy* (Boston: Beacon Press, 1984), 1.

14. Laurence Bergreen, *Columbus: The Four Voyages* (New York: Viking, 2011).

15. Anne McClintock, *Imperial Leather: Race, Gender and Sexuality in the Colonial Context* (New York: Routledge, 1995), 21–24, esp. 22.

16. Deborah A. Miranda, "Extermination of the Joyas: Gendercide in Spanish California," *GLQ: A Journal of Lesbian and Gay Studies* 16, no. 1–2 (2010): 253–284.

17. Andrea Smith, *Conquest: Sexual Violence and American Indian Genocide* (Durham, NC: Duke University Press, 2015).

18. Merchant, *Death of Nature*, 1980; Evelyn Fox Keller, *Reflections on Gender and Science* (New Haven, CT: Yale University Press, 1985).

19. Jane Caputi, "A (Bad) Habit of Thinking: Challenging and Changing the Pornographic Worldview," in *Women in Popular Culture: Representation and Meaning*, ed. Marian Meyers (Cresskill, NJ: Hampton Press, 2008), 29–56. See also Caputi, "Dirt," in *Encyclopedia of Religion and Nature*, ed. Bron R. Taylor (New York: Thoemmes Continuum, 2005), 486–487.

20. O'Brien suggests substituting animality for *disability* as a way of registering this reality. Ruth O'Brien, *Bodies in Revolt: Gender, Disability, and a Workplace Ethic of Care* (New York: Routledge, 2005), 5–6.

21. Vanessa Watts, "Indigenous Place-Thought and Agency amongst Humans and Non-Humans (First Woman and Sky Woman Go on a European World Tour!)," *Decolonization: Indigeneity, Education and Society* 2, no. 1 (2013): 20–34, esp. 27.

22. Susan Griffin, "Split Cultures," in *Healing the Wounds: The Promise of Ecofeminism*, ed. Judith Plant (Philadelphia: New Society, 1989), 7–17; Williams, "Sin, Nature, and Black Women's Bodies"; Gloria Anzaldúa, *Borderlands/ La Frontera: The New Mestiza* (San Francisco: Spinsters /Aunt Lute Press, 1987); Vandana Shiva, *Staying Alive: Women, Ecology and Survival in India* (London: Zed Books, 1988); Plumwood, *Feminism and the Mastery of Nature*; Mies, *Patriarchy and Accumulation*; Greta Gaard, "Toward a Queer Ecofeminism," *Hypatia* 12, no. 1 (1997): 114–137; Sjöö and Mor, *Great Cosmic Mother*; Donna Haraway, *The Companion Species Manifesto: Dogs, People, and Significant Otherness* (Chicago: Prickly Paradigm Press, 2003).

23. Sjöö and Mor, *Great Cosmic Mother*, 21.

24. Sterling Stuckey, *Slave Culture: Nationalist Theory and the Foundations of Black America*, 25th anniversary ed. (New York: Oxford University Press, 1987, 2013), 25.

25. Serial sex killers sometimes speak of "cleansing" their sex worker victims. Jane Caputi, "Gods We Worship: Sexual Murder

as Religious Sacrifice," in *Goddesses and* Monsters, 182–206, esp.186–187.

26. Audre Lorde, ""Uses of the Erotic: The Erotic as Power," *Sister Outsider* (Trumansburg, NY: Crossing Press, 1984), 53–59; Elias Farajajé-Jones, "Holy Fuck," in *Male Lust: Pleasure, Power, and Transformation*, ed. Kerwin Kay, Jill Nagle, and Baruch Gould (New York, Harrington Park Press, 2000), 327–336; Gaard, "Toward a Queer Ecofeminism."

27. Gayle S. Rubin, "Thinking Sex: Notes for a Radical Theory of the Politics of Sexuality (1984)," *Deviations: A Gayle Rubin Reader* (Durham, NC: Duke University Press, 2011), 137–181.

28. farajajé-jones, "Holy Fuck," 328–331.

29. farajajé-jones, "Holy Fuck," 330.

30. Greg Cajete, *Native Science: Natural Laws of Interdependence* (Santa Fe, NM: Clear Light, 2000), 153.

31. Louise Erdrich, *The Last Report on the Miracles at Little No Horse* (New York: HarperCollins, 2001), 134.

32. Quoted in Robert May, *Sex and Fantasy* (New York: W. W. Norton, 1980), 140. This is from John Updike, "London Life," *Picked-Up Pieces* (New York: Alfred A. Knopf, 1975).

33. Dworkin, "The Root Cause," in *Our Blood: Prophecies and Discourses on Sexual Politics*, 96–111, esp. 105–107.

34. Jane Caputi, *Gossips, Gorgons and Crones: The Fates of the Earth* (Santa Fe: Bear & Company, 1993).

35. The *Me Too* website features a statement by founder Tarana Burke and a brief history, https://metoomvmt.org/. See also the *Times Up* movement, which works against sexual harassment. https://www.timesupnow.com/.

36. Tina Ngata, "Sex, Filth, Violence and the Pacific," June 30, 2018, *The Non-Plastic Maori* (blog), https://thenonplasticmaori.word-press.com/author/tinangata/.

37. Lorde, "Uses of the Erotic, 53.

38. Samy H. Alim and Geneva Smitherman, *Articulate While Black: Barack Obama, Language, and Race in America* (New York: Oxford University Press, 2012), 122.

39. Sharpe, *In the Wake*, 15.

40. For example, a 1984 political cartoon by David Levine, "Screwing the World," shows a grinning and self-satisfied Henry Kissinger

draped in the American flag and missionary-position fucking a woman whose head is planet Earth, https://illustrationchronicles.com/Screwing-the-World-with-David-Levine. In May 2018, Samantha Bee updated the imprecation, aiming it at Scott Pruitt, the fossil-fuel enthusiast appointed by Donald Trump to direct the Environmental Protection Agency. He is shown holding Earth in his hands and seeming just about ready to begin thrusting. Bee comments: "Who says Scott Pruitt doesn't love the Earth?" Samantha Bee, https://twitter.com/FullFrontalSamB/status/994409637633052673.

41. June Jordan, "Getting Down to Get Over," in *Directed by Desire: The Collected Poems of June Jordan*, ed. Jan Heller Levi and Sara Miles (Port Townsend, WA: Copper Canyon Press, 2005), 172–183, esp. 174–175.

42. André 3000 (André Benjamin) with OutKast, "Vibrate," recorded in 2003, track 19 on *The Love Below*, part of *Speakerboxxx/The Love Below*, LaFace Records, compact disc, © BMG Rights Management.

43. *Strong language* is "forceful or offensive language, esp. used as an expression of anger or strong feeling" (*OED*).

44. Alim and Smitherman, *Articulate While Black*, 109.

45. James Brown, *Mutha's Nature*, recorded in 1977, Polydor, compact disc. Brown introduces his cover of the song written by D. Hayward and George Gershwin, "Summertime," track 6.

46. House of Downtown, "Paradise," lyrics by E. Todd and C. Erdolano, A. McCarthy, on *Mutha Funkin Earth*, 2003, Universal Music of New Zealand, track 1, compact disc.

47. Jane Caputi, *The Age of Sex Crime* (Bowling Green, OH: Bowling Green State University Press, 1987), 192.

48. L. H. Stallings, *Mutha' Is Half a Word: Intersections of Folklore, Vernacular, Myth, and Queerness in Black Female Culture* (Columbus: Ohio State University Press, 2007), 10–11.

49. Brittany Spanos, "Janelle Monáe Frees Herself," *Rolling Stone*, April 26, 2018, https://www.rollingstone.com/music/music-features/janelle-monae-frees-herself-629204/.

50. Stallings, *Mutha' Is Half a Word*, 25.

51. Wynter, Sylvia. "Sambos and Minstrels," *Social Text*, no. 1 (1979): 149–156.

52. Earth worker Amber Tamm hails "Momma Earth's wisdom" as teaching her "how to self generate." https://www.instagram.com/p/B0Yn9CpFvqn/.

53. Paula Gunn Allen argues that contemporary movements for green, women's and LGBTQ rights and justice derive from spirit-based way of being and doing developed by Indigenous peoples. Paula Gunn Allen, *The Sacred Hoop: Recovering the Feminine in American Indian Traditions* (Boston: Beacon Press, 1986), 114–115.

54. Patrisse Cullors and Nyeusi Nguvu, "From Africa to the US to Haiti, Climate Change Is a Race Issue," *The Guardian*, Sept. 14, 2017,

55. Hebah H. Farrag, "The Role of Spirit in the BlackLivesMatter Movement: A Conversation with Activist and Artist Patrisse Cullors," *Religion Dispatches*, June 24, 2015, http://religiondispatches.org/the-role-of-spirit-in-the-blacklivesmatter-movement-a-conversation-with-activist-and-artist-patrisse-cullors/.

56. Gail Dillon alludes to Archer Pechawis's (Cree) short film *Horse*. See introduction to *Walking the Clouds: An Anthology of Indigenous Science Fiction*, ed. Grace L. Dillon (Tucson: University of Arizona Press, 2012), 5.

57. Avinash Chak, "Beyond 'He' and "She': The Rise of Non-Binary Pronouns," BBC News, Dec. 7, 2015, https://www.bbc.com/news/magazine-34901704.

58. *Ask a Non-Binary* (blog), Nov. 9, 2013, http://askanonbinary.tumblr.com/post/66494772838/pronouns-i-have-encountered-in-no-particular-order.

59. On volatile essences, see Jane Caputi, "Cunctipotence: Elemental Female Potency," *Trivia: A Journal of Ideas*, no. 4 (Summer 2006), https://www.triviavoices.com/cunctipotence.html.

60. Liza Cowan's full statement is included on the Etsy website, https://www.smallequals.com/listing/274863796/mother-nature-is-a-lesbian-feminist.

61. *Wigstock: The Movie*, dir. Barry Shils (1995, Samuel Goldwyn), film.

62. Elizabeth M. Stephens and Annie M Sprinkle, "Ecosex Manifesto," SexEcology website, n.d., http://sexecology.org/research-writing/ecosex-manifesto/.

63. See the Commonality Institute, based on the work of poet and activist Judy Grahn, https://commonalityinstitute.org/.
64. Amir Khadar, "Trans Day of Resilience Poster," created as a commission from Forward Together for Their Trans Day of Resilience Celebration, 2017. The Trans Day of Resilience, founded with the Audre Lorde Project, is November 20. It is an addition to the "Trans Day of Remembrance" of those murdered in hate crimes. For more, see https://tdor.co/.
65. Marilou Awiakta, *Selu: Seeking the Corn-Mother's Wisdom* (Golden, CO: Fulcrum, 1993), 6
66. Carolyn Merchant, *Autonomous Nature: Problems of Prediction and Control from Ancient Times to the Scientific Revolution* (New York: Routledge, 2016).
67. Claire Henry, *Revisionist Rape Revenge: Redefining a Film Genre* (New York: Palgrave Macmillan, 2014).
68. *Maamé* is a "word in Akan . . . for the Great Mother, the maternal spirit in all things." Alvia J. Wardlaw, "Metamorphosis: The Life and Art of John Biggers," in *The Art of John Biggers: View from the Upper Room* (Houston, TX: Museum of Fine Arts Houston, 1995), 47.
69. The names *Mother Love, Big Mama, Ma Dear, Earth Mother* are all invoked by Susan L. Taylor, "The Goddess Within," *Essence*, October 2006, 250.
70. "Big Mama Nature," *Fiyah: Magazine of Black Speculative Fiction*, issue 6, Spring 2018, https://www.fiyahlitmag.com/product/issue-6-big-mama-nature/.
71. Cherríe Moraga writes, "*The earth is female.* Whether myth, metaphor, or memory, she is called 'Mother' by all peoples of all time. *Madre Tierra.*" Cherríe Moraga, *The Last Generation: Prose and Poetry* (Boston: South End Press, 1993), 172.
72. "The Indweller of Earth" of the Chugach Inuit is "understood to be a woman, her body encased in a bright light: she wears a garment from which are suspended all the terrestrial creatures. She dwells in the mountain forests." Jordan Paper, *Through the Earth Darkly: Female Spirituality in Comparative Perspective* (New York: Continuum, 1997), 117.
73. Emily Culpepper, "Gorgons: A Face for Contemporary Women's Rage," *Woman of Power* 3 (1986): 22–24, 40. See also Caputi, *Gossips, Gorgons and Crones.*

74. Hine is a Maori deity, the one who births all into being and is also the bringer of death. Tina Ngata, "I am Hine, I am Moana," YouTube video, Nov. 28, 2016, https://www.youtube.com/watch?v=f2b5TRv23RQ&list=UUeYfTjLXSWTupnusvp3W6Og&index=12&t=0s. This video features artworks by Robyn Kahukiwa.

75. The artist Amalia Mesa Bains gives "Cihuateotle" as an ancient name of "mother nature" in her moss-covered sculpture "Cihuateotl (Woman of Cihuatlampa)," discussed in Christina Holmes, *Ecological Borderlands: Body, Nature, and Spirit in Chicana Feminism* (Urbana: University of Illinois Press, 2013), 88–91.

76. Taino Earth Mother. Patricia Monaghan, *The Book of Goddesses and Heroines*, rev. and enlarged ed. (St. Paul, MN: Llewellyn, 1990), 38–39.

77. Sila, an Inuit understanding of the primal force-source, is not gendered. Zoe Todd writes: "The infinitesimal bit of the concept of Sila that I can claim to understand is that it is bound with life, with climate, with knowing, and with the very existence of being(s). And, in some respects, it sounds an awful lot like the idea of Gaia to my Métis ears." Todd, "Indigenous Feminist's Take on the Ontological Turn," 5.

78. Judith Gleason, *Oya: In Praise of the Goddess* (Boston: Shambhala, 1987).

79. Kali, "the Primal Mother who brings forth all life even while she signifies Death." Vrinda Dalmiya, "Loving Paradoxes: A Feminist Reclamation of the Goddess Kali," *Hypatia* 15, no. 1 (2000): 125–150, esp.136.

80. *Amalur* is the Basque name of Mother Earth.

81. Diné (Navajo) "Mother Earth." Esther Yazzie-Lewis and Jim Zion, "Leetso, the Powerful Yellow Monster: A Navajo Cultural Interpretation of Uranium Mining," in *The Navajo People and Uranium Mining*, ed. Doug Brugge, Timothy Benally, and Esther Yazzie (Albuquerque: University of New Mexico Press, 2007), 1–10, esp. 6.

82. *Nüwa* is the Chinese creator, originally both earth and sky, female-identified, containing the masculine, later split into a brother-sister pair. Paper, *Through the Earth Darkly*, 55.

83. Garrard, *Brunelleschi's Egg*, 10.

84. "A female deity associated with the powers of the land in the religious traditions of the Igbo of southeastern Nigeria," Robert M. Baum, "Ala," in *Encyclopedia of Women and World Religion*, ed. Serenity Young, vol. 1 (New York: Macmillan, 1999), 22.

85. Muzzu-Kummik-Quae is "Earth Mother" in Anishinaubae cultures. Basil Johnston, *The Manitous: The Spiritual World of the Ojibway* (New York: HarperCollins, 1995), 243.

86. *Bhumi*, or "Existing" is one name of Mother Earth in India. Martha Ann and Dorothy Myers Imel, *Goddesses in World Mythology* (Santa Barbara, Calif.: ABC-CLIO, 1993), 252.

87. Hawai'ian Earth Mother. See also Meghan Miner Murray, "Why Are Native Hawaiians Protesting against a Telescope?," *New York Times*, July 22, 2019.

88. *Unci Maka* is Mother Earth in Lakota. Ladonna Bravebull Allard, "To Save the Water, We Must Break the Cycle of Colonial Trauma," *Camp of the Sacred Stones* (blog), Feb. 24, 2017, http://sacredstonecamp.org/blog/2017/2/4/to-save-the-water-we-must-break-the-cycle-of-colonial-trauma.

89. Irene Lara, "Sensing the Serpent in the Mother, *Dando a Luz la Madre Serpiente*: Chicana Spirituality, Sexuality, and Mamihood," in *Fleshing the Spirit: Spirituality and Activism in Chicana, Latina, and Indigenous Women's Lives*, ed. Elisa Facio and Irene Lara (Tucson: University of Arizona Press, 2014), 113–134, esp.115.

90. *Inang Kalikasan* is the Philippine Mother Nature. "Nature Is Speaking," Conservation International Philippines, n.d., https://www.conservation.org/global/philippines/Pages/Inang-Kalikasan.aspx.

Prelude to a Curse

1. "Magic Words," in *Shaking the Pumpkin: Traditional Poetry of the Indian North Americas*, ed. J. Rothenberg (Albuquerque: University of New Mexico Press, 1986).

2. Stephanie Hayes, "Curses! How to Get Ahead by Swearing," *Atlantic*, January–February, 2016, 8.

3. In a conversation with me, Toni Francis urged me to stress the paradigmatic character of motherfucking. Boca Raton, December, 2016.

Chapter 3

1. Carson, "Of Man and the Stream of Time," 426.
2. Rachel Carson, "Preface to Animal Machines" (1964), in *Lost Woods: The Discovered Writing of Rachel Carson*, ed. Linda Lear (Boston: Beacon Press, 1999), 194–196, esp. 194.
3. W. Steffen, A. Sanderson, P. D. Tyson, J. Jäger, P. A. Matson, B. Moore III, F. Oldfield, K. Richardson, H. J. Schellnhuber, B. L. Turner, and R. J. Wasson, "Great Acceleration, Global Change, International Geosphere-Biosphere-Programme," adapted from *Global Change and the Earth System: A Planet under Pressure* (New York: Springer-Verlag, 2004), http://www.igbp.net/globalchange/greatacceleration.4.1b8ae20512db692f2a680001630.html. See also Will Steffen, Wendy Broadgate, Lisa Deutsch, "The Trajectory of the Anthropocene: The Great Acceleration," *Anthropocene Review* 2, no. 1 (2015): 81–98, https://doi.org/10.1177/2053019614564785.
4. Susan Wendell, *The Rejected Body: Feminist Philosophical Reflections on Disability* (New York: Routledge, 1996), 59–60.
5. Gabrielle Hecht, "Residue," *Somatosphere*, Jan. 8. 2018, http://somatosphere.net/2018/01/residue.html.
6. Diane Ackerman, *The Human Age: The World Shaped by Us* (New York: W. W. Norton, 2014), 307.
7. David Grinspoon, *The Earth in Human Hands: Shaping Our Planet's Future* (New York: Grand Central, 2016), ix.
8. Grinspoon, *Earth in Human Hands*, xvii, xiv, 453, xiv, 453.
9. Sarah Deer, *The Beginning and End of Rape: Confronting Sexual Violence in Native America* (Minneapolis: University of Minnesota Press, 2015), xix.
10. Ackerman, *Human Age*, 10.
11. Ackerman, *Human Age*, 11.
12. Ackerman, *Human Age*, 308.
13. Grinspoon, *Earth in Human Hands*, 165.
14. Owen Gaffney and Félix Pharand-Deschênes, *Welcome to the Anthropocene*. 2012, length: 3:38 min, video. https://vimeo.com/39048998. The words are from the voiceover, http://www.environmentandsociety.org/mml/welcome-anthropocene.

15. Gabrielle Hecht, "The African Anthropocene," *Aeon*, November 2013, https://aeon.co/users/gabrielle-hecht.

16. These seventeen principles were formulated Oct. 21–24, 1991, in Washington DC by delegates to the First National People of Color Environmental Leadership Summit, https://www.ejnet.org/ej/principles.html.

17. For more on such hoarding, see Shatema Threadcraft, *Intimate Justice: The Black Female Body and the Body Politics* (New York: Oxford University Press, 2016), 152–166.

18. Peggy V. Beck, Anna Lee Walters, and Nia Francisco, *The Sacred: Ways of Knowledge, Sources of Life*, redesigned ed. (Tsaile, AZ: Navajo Community College Press, 1996), 6.

19. This is included in the *Oxford English Dictionary* definition of *worship*.

20. Crutzen and Schwägerl, "Living in the Anthropocene."

21. Rene Descartes, "Discourse on Method Part 4," in *Philosophical Works of Descartes*, ed. E. S. Haldane and G. R. T. Ross (New York: Dover, 1955), vol. 1, p. 119.

22. Slavoj Žižek, "Ecology against Mother Nature: Slavoj Žižek on Molecular Red," *Verso*, May 26, 2015, http://zizek.uk/ecology-against-mother-nature-slavoj-zizek-on-molecular-red/3/.

23. Merchant, *Death of Nature*, 171. Francis Bacon used the term "common harlot."

24. Carolyn Merchant, "Science and Worldviews," in *Radical Ecology: The Search for a Livable World* (New York: Routledge, 1992), 41–60, esp. 46.

25. Merchant, *Death of Nature*, 189.

26. Merchant, *Death of Nature*, 190.

27. Spencer Weart, *Nuclear Fear: A History of Images* (Cambridge, MA: Harvard University Press, 1989), 58.

28. Jennifer Doudna, *A Crack in Creation: Gene Editing and the Unthinkable Power to Control Creation* (Boston: Houghton Mifflin Harcourt, 2017), 236.

29. White, "The Historical Roots of Our Ecological Crisis," 197.

30. Daly and Caputi, *Webster's First New Intergalactic Wickedary*, 195.

31. Vanessa Watts, "Indigenous Place-Thought and Agency amongst Humans and Non-Humans (First Woman and Sky Woman Go on a European World Tour!)," *Decolonization, Indigeneity,*

Education and Society 2, no. 1 (2013): 20–34, esp. 20, http://www. decolonization.org/index.php/des/article/view/19145.

32. Mary Daly, *Pure Lust: Elemental Feminist Philosophy* (Boston: Beacon Press, 1984), 7–20. See also Kim Power, *Veiled Desire: Augustine on Women* (New York: Continuum, 1996), 53.

33. White, "Historical Roots," 197.

34. Crutzen and Schwägerl, "Living in the Anthropocene."

35. Andrews, *Dictionary of Nature Myths*, 61.

36. N. Scott Momaday, *The Way to Rainy Mountain* (Albuquerque: University of New Mexico Press, 1969), 3, 8–10.

37. Yuval Noah Harari, *Homo Deus: A Brief History of Tomorrow* (New York: HarperCollins, 2017), 71.

38. Caputi, *Age of Sex Crime.*

39. Quoted in "The Random Killers," *Newsweek*, Nov. 26, 1984, 100–106D, esp. 104.

40. Bill Pennington, "Josh Brown's Past Admission of Domestic Abuse Causes N.F.L. to Reopen Inquiry," *New York Times*, Oct. 20, 2016.

41. Peter Langman, "Columbine, Bullying and the Mind of Eric Harris," *Psychology Today*, May 20, 2009, https://www.psychologytoday.com/us/blog/keeping-kids-safe/200905/columbine-bullying-and-the-mind-eric-harris.

42. Brian Dakss, "Sniper Note 'Call Me God,'" CBS News, October 26, 2002, https://www.cbsnews.com/news/sniper-note-call-me-god/.

43. Linda Massarella, Sophia Rosenbaum, and Leonard Greene, "The Vile Manifesto of a Killer," *New York Post*, May 25, 2014.

44. Mark Baker, *Nam: The Vietnam War in the Words of the Men and Women Who Fought There* (New York: William Morrow, 1982), 152.

45. Edmund Leach, *A Runaway World?* (New York: Oxford University Press, 1968), 1.

46. John Brockman, "We Are as Gods and Have to Get Good at It: Steward Brand Talks about His Ecopragmatist Manifesto," *Edge*, Aug. 18, 2009, https://www.edge.org/conversation/stewart_brand-we-are-as-gods-and-have-to-get-good-at-it.

47. This is the description on Amazon.com for Lynas, *The God Species*, https://www.amazon.com/God-Species-Saving-Planet-Humans/dp/142620891X.

48. Kathleen Miles, "Ray Kurzweil: In the 2030s, Nanobots in Our Brains Will Make Us 'Godlike,'" *Huffington Post*, Oct. 26, 2017. https://www.huffingtonpost.com/entry/ray-kurzweil-nanobots-brain-godlike_us_560555a0e4b0af3706dbe1e2.
49. Harari, *Homo Deus*, 43.
50. "He's Got the Whole World in His Hands," n.d., https://hymnary.org/text/hes_got_the_whole_world_in_his_hands.
51. This page is from the *Ted Radio Hour*, "The Anthropocene," September 30, 2016, https://www.npr.org/programs/ted-radio-hour/494774287/anthropocene.
52. Daly, *Pure Lust*, 10.
53. I discuss this in depth in Gossips, Gorgons, and Crones.
54. "Father God Created Mother Earth," no designer specified, Zazzle, https://www.zazzle.com/father_god_created_mother_earth_postcard-239607089276208815).
55. Miniature from Bible moralisée, Reims, France, National Library of Austria, Vienna, COD. 2554, fol, 1v. Photo: Erich Lessing/Art Resource, NY. Garrard, *Brunelleschi's Egg*, 2.
56. Garrard, *Brunelleschi's Egg*, 13.
57. I first saw this in Tony Horwitz, *A Voyage Long and Strange: Rediscovering the New World* (New York: Henry Holt, 2008), 115.
58. The Papal Bull "Inter Caetera," issued by Pope Alexander VI on May 4, 1493, "Doctrine of Discovery 1493," Gilder Lehrman Institute of American History, https://www.gilderlehrman.org/content/doctrine-discovery-1493.
59. *The Economist*, May 28–June 3, 2011, https://www.economist.com/leaders/2011/05/26/welcome-to-the-anthropocene.
60. Kevin Kelly, *What Technology Wants* (New York: Viking Press 2010), 12.
61. Kelly, *What Technology Wants*, 325.
62. Tom Bartlett, "What Separates Man From Beast? Creativity," *Chronicle of Higher Education*, Dec. 10, 2017.
63. Natalie Angier, "Termites: Guardians of the Soil," *New York Times*, March 2, 2015. JoAnna Klein. "What Termites Can Teach Us about Cooling Our Buildings," *New York Times*, March 26, 2019.
64. Kolbert, "Enter the Anthropocene," 2011.

65. Robert Mankoff, "Now That's Product Placement." *New Yorker,* Nov. 22, 1999, http://www.condenaststore.com/-sp/Now-that-s-product-placement-New-Yorker-Cartoon-Prints_i8474558_.htm.

66. Garrard, *Brunelleschi's Egg,* 49.

67. Plumwood, *Feminism and the Mastery of Nature,* 93.

68. Jonathan Shaw, "Skyscraper as Symbol: The Semiotics of Skyscrapers," *Harvard Magazine,* May–June 2010, https://harvardmagazine.com/2010/05/skyscraper-as-symbol.

69. Sarah Gamble, *The Routledge Critical Dictionary of Feminism and Postfeminism* (New York: Routledge, 1999), 294.

70. Harari, *Homo Deus,* 47.

71. Jeffrey J. Kripal, "Phallus and Vagina," *Encyclopedia of Religion,* ed. Lindsay Jones. 2nd ed, vol. 10 (Detroit: Macmillan Reference USA, 2005), 7077–7086, esp. 7077.

72. Daly, *Pure Lust,* 226.

73. Susan Bordo, *The Male Body: A New Look at Men in Public and in Private* (New York: Farrar Straus and Giroux, 1999), 89.

74. Scott Poulson-Bryant, *Hung: A Meditation of the Measure of Black Men in America* (New York: Doubleday, 2005).

75. Ann Baring and Jules Cashford, *The Myth of the Goddess: Evolution of an Image* (New York: Arkana, 1991), 275.

76. Michelle I. Marcus, "Sex and the Politics of Female Adornment in Pre-Achaemenid Iran (1000–800 B.C.E)," in *Sexuality in Ancient Art: Near East, Egypt, Greece, and Italy,* ed. Natalie Boymel Kampen (Cambridge: Cambridge University Press, 1996), 50.

77. Baring and Cashford, *Myth of the Goddess,* 275.

78. *Scientific American,* Dec. 2015, https://www.scientificamerican.com/article/readers-respond-to-how-we-conquered-the-planet/.

79. *Feed the Green: Feminist Voices for the Earth,* a film by Jane Caputi (2015, distributed by Women Make Movies), DVD, http://www.wmm.com/filmcatalog/pages/c934.shtml.

80. I first saw this photo in Sarah Hrdy, *Mother Nature: Natural Selection and the Female of the Species* (London: Catto & Windus, 1999), 426. Grancel Fitz, *Big Baby,* Keith de Lellis Gallery, New York.

81. Teresa Brennan, *History after Lacan* (New York: Routledge, 1993), 167.

82. *Feed the Green*.

83. ©2005 Ford Motor Company.

84. Henry Hook, Emily Cox, and Henry Rathvon, *The Boston Globe Sunday Crossword Omnibus*, vol. 2 (New York: Random House, 2003). The cover was designed by June Chan and illustrated by Stuart Bragg.

85. Allen, *Sacred Hoop*, 119.

86. Lynas, *God Species*, 5.

87. Neil DeGrasse Tyson, interview, *On Point*, National Public Radio, May 2, 2017.

88. Church and Regis, *Regenesis*, 143.

89. Church and Regis, *Regenesis*, 2.

90. Cited in Howard Zinn, *A People's History of the United States* (New York: Harper Perennial Modern Classics, 2015), 1.

91. Mary Daly, *Quintessence: Realizing the Archaic Future A Radical Elemental Feminist Manifesto* (Boston: Beacon Press, 1998), 222.

92. Ira Levin, *The Stepford Wives* (New York: Random House, 1972).

93. Mary Daly, *Gyn/Ecology: The Metaethics of Radical Feminism* (Boston: Beacon Press, 1978), 70–71.

94. Tuana, *Less Noble Sex*, xi.

95. Hurston, *Their Eyes Were Watching God*.

96. Garrard, *Brunelleschi's Egg*, 54.

97. Hans Moravec, *Mind Children: The Future of Robot and Human Intelligence* (Cambridge, MA: Harvard University Press, 1988).

98. Kurzweil cited in Ackerman, *Human Age*, 181.

99. Alfred Kubin, *Unser aller Mutter Erde* (*Earth: Mother of Us All*), 1900, ink on paper, 6 3/4 x 15". The illustration is viewable at Karen Rosenberg, "Mapping the Shadowy Corners of the Subconscious," *New York Times*, Oct. 15, 2008, https://www.nytimes.com/2008/10/16/arts/design/16neue.html.

100. Max More, "A Letter to Mother Nature," in *The Transhumanist Reader*, ed. Max More and Natasha Vita-More (Malden, MA: Wiley-Blackwell, 2013), 449–450.

101. More, "Letter to Mother Nature," 449–450.

102. More, "Letter to Mother Nature," 450.

103. Robert Pogue Harrison, *Forests: The Shadow of Civilization* (Chicago: University of Chicago Press, 1992), 14.

104. Harrison, *Forests*, 14.

105. Harrison, *Forests*, 14.

106. Ann Barnard, "Climate Change Is Killing the Cedars of Lebanon," *New York Times*, July 18, 2018.

107. Nikola Danaylov, "A Transhumanist Manifesto [Redux]," *Singularity Weblog*, March 11, 2016, https://www.singularity-weblog.com/a-transhumanist-manifesto/.

108. Nocella II, Bentley, Duncan, "Introduction," xvi–xvii.

109. Benjamin Reeves, "The Search for Immortality," *Worth: The Evolution of Financial Intelligence*, April–May 2015, 061–067.

110. "Ecological Footprint Calculator," http://www.footprintcalculator.org/.

111. The World Health Organization estimates that nine million people die prematurely every year from pollution, the majority because of air pollution. Though it is not usually mentioned, locations in the so-called developing nations suffer the brunt of this. See http://time.com/4989641/water-air-pollution-deaths/

112. This photo is ©Karine Aigner for a card distributed and sold exclusively by Trader Joe's.

113. Stanley Bing, "The Downloadable Brain: We're Closer Than We Think to Immortality," *Literary Hub*, Dec. 8, 2017, https://lithub.com/the-downloadable-brain-were-closer-than-we-think-to-immortality/.

114. *Get Out*, written and dir. by Jordan Peele (2017; Universal Pictures), film.

115. Memphis Slim, "Mother Earth," 1951, Buddah Records, *Mother Earth*, track 12, 1972, compact disc.

116. Favianna Rodriguez, "Defiende Nuestra Madre," 2014. http://favianna.flyingcart.com/index.php?p=detail&pid=233&cat_id=.

117. Shiva, *Staying Alive*, 41–42.

118. Shiva, interviewed in *Feed the Green*.

119. Tonya Gonnella Frichner, "Mother Earth Is a Relative, Not a Resource," *Indian Country Today*, April 22, 2014.

120. Leroy Little Bear, "Jagged Worldviews Colliding," in *Reclaiming Indigenous Voice and Vision*, ed. Marie Battiste (Vancouver: University of British Columbia Press, 2000), 77–85, esp. 78.

121. Marilou Awiakta, telephone conversation with the author, Jan. 28, 2017.

122. "Preamble," *Rights of Mother Earth*, 2010, Alliance for Democracy, http://www.thealliancefordemocracy.org/pdf/AfDJR6117.pdf.

123. J. Vidal, "Bolivia Enshrines Natural World's Rights with Equal Status for Mother Earth," *The Guardian*. April 10, 2011.

124. Eleanor Ange Roy, "New Zealand River Granted Same Legal Rights as Human Being," *The Guardian*, March 16, 2017.

125. Kim TallBear, "Badass Indigenous Women Caretake Relations #STANDINGROCK, #IDLENOMORE, #BLACKLIVESMATTER," in *Standing with Standing Rock*, ed. Nick Estes and Jaskiran Dhillon (Minneapolis: University of Minnesota Press, 2019), 13–18.

126. Cited in TallBear, "Badass Indigenous Women Caretake Relations," 13.

127. "On the Anniversary of Sitting Bull's Death, Meet His Great Great Granddaughter, Brenda White Bull," *Indigenous Environmental Network*, n.d.,http://www.ienearth.org/on-the-anniversary-of-sitting-bulls-death-meet-his-great-great-granddaughter-brenda-white-bull/.

128. Jaskiran Dhillon and Nick Estes, "Introduction: Standing Rock, #NoDAPL, and Mni Wiconi," *Cultural Anthropology*, Dec. 22, 2016.

129. Egla Martínez Salazar, email correspondence, Feb. 21, 2017 and Dec. 22, 2018. See also Egla Martínez Salazar, *Global Coloniality of Power in Guatemala: Racism, Genocide, Citizenship* (New York: Lexington Books, 2014).

130. Nina Lakhani, "Berta Cáceres: Conviction of Killers Brings Some Justice, but Questions Remain," *The Guardian*, Dec. 1, 2018.

131. Jonathan Watts and John Vidal, "Environmental Defenders Being Killed in Record Numbers Globally, New Research Reveals," *The Guardian*, July 13, 2017.

132. Joseph Nevins, "How US Policy in Honduras Set the Stage for Today's Migration," *The Conversation*, October 31, 2016, http://theconversation.com/how-us-policy-in-honduras-set-the-stage-for-todays-migration-6593. Egla Martínez Salazar told me of Caceres's use of this phrase, email correspondence Dec. 21, 2018.

133. Grahame Russell, "Berta Cáceres: Who She Is & What She Lived For," *Rights Action*, March 2, 2018, https://mailchi.mp/rightsaction/berta-caceres-who-she-is-what-she-lived-for.

134. "Who We Are," COPINH, http://copinhenglish.blogspot.com/p/who-we-are.html.

135. Berta Cáceres, acceptance speech, 2015 Goldman Prize ceremony, YouTube, https://www.youtube.com/watch?v=AR1kwx8b0ms.

136. Jason Wilson, "Eco-Fascism Is Undergoing a Revival in the Fetid Culture of the Extreme Right," *The Guardian*, March 19, 2019.

137. Nocella II, Bentley and Duncan, "Rise of Eco-Ability," xvii.

138. Watts, "Indigenous Place-Thought & Agency," 20.

139. Watts, "Indigenous Place-Thought & Agency," 30–31.

140. Paula Gunn Allen, *Sacred Hoop*, 268.

141. Paula Gunn Allen, *Grandmothers of the Light: A Medicine Woman's Sourcebook* (Boston: Beacon Press, 1991), 108.

142. Watts, "Indigenous Place-Thought & Agency," 32.

143. Watts, "Indigenous Place-Thought & Agency," 25.

144. I am revising a statement I made in *The Age of Sex Crime*, 9.

145. Jane Caputi, "Shifting the Shapes of Things to Come: The Presence of the Future in The Work of Gloria Anzaldúa," in *Entre Mundos/ Among Worlds: New Perspectives on Gloria* Anzaldúa, ed. AnaLouise Keating (New York: Palgrave Macmillan, 2005), 185–194.

146. Gaspar Sánchez of COPINH spoke in Charlestown, Massachusetts, on Nov. 9, 2017 as part of a Witness for Peace Tour. He was wearing a T-shirt with this same image and told me (in Spanish and then translated) that it had been composed "by the collective."

147. This translation was given to me by Dr. Mary Ann Gosser Esquilín, Professor, Department of Language and Linguistics, Florida Atlantic University.

148. Matt Ginsberg-Jaeckle, "Planting Berta: A Year of Impunity, a Life of Rebellion and the Roots of a New World," *Honduras Resists*, March 9, 2017, http://hondurasresists.blogspot.com/2017/03/planting-berta-year-of-impunity-life-of.html.

149. Carolyn Ziv, "'She Is a Seed That Has Multiplied': How the Murder of Berta Cáceres Launched a Movement," *Fund for*

Global Human Rights, May 29, 2018, https://globalhumanrights. org/she-is-a-seed-that-has-multiplied-how-the-murder-of-berta-caceres-launched-a-movement/.

150. Luther Standing Bear, "Nature (1933)," in Gottlieb, *This Sacred Earth*, 39–42.

151. Faith Spotted Eagle, quoted in Native Youth Sexual Health Network and Women's Earth Alliance, *Violence on the Land, Violence on Our Bodies*, 48, http://landbodydefense.org/uploads/files/VLVBReportToolkit2016.pdfHealt.

152. Angela Y. Davis, "Masked Racism: Reflections on the Prison Industrial Complex," *Colorlines*, Sept. 10, 1998, https://www.colorlines.com/articles/masked-racism-reflections-prison-industrial-complex. Beth E. Richie, *Arrested Justice: Black Women, Violence, and America's Prison Nation* (New York: New York University Press, 2012); Lisa A. Kemmerer, ed, *Sister Species: Women, Animals and Social Justice* (Urbana: University of Illinois Press, 2011); Kemmerer, *Eating Earth: Environmental Ethics and Dietary Choice* (New York: Oxford University Press, 2014).

153. Erin Marie Konsmo, "Land Is Ceremony," in *Violence on the Land, Violence on Our Bodies*, 10.

154. Robin Wall Kimmerer, *Braiding Sweetgrass* (Minneapolis, MN: Milkweed Editions, 2013), 5.

155. Wallace Black Elk and William S. Lyon, *The Sacred Ways of a Lakota* (San Francisco: HarperSanFrancisco, 1990), 187.

156. This is part of a statement made by a Native speaker (unidentified) speaking against fossil fuel extraction on the Northern Cheyenne reservation in Montana in the documentary, *This Changes Everything*, dir. by Avi Lewis (2015; Canada, Klein-Lewis Productions) film.

157. Harriet McBryde Johnson speaks to the pleasures of rolling in *Too Late to Die Young* (New York: Henry Holt, 2005), 250–261. See also Alison Kafer, *Feminist Queer Crip* (Bloomington: Indiana University Press, 2013).

158. Eve Tuck and K. Wayne Yang, "Decolonization Is Not a Metaphor," *Decolonization: Indigeneity, Education and Society* 1, no. 1 (2012): 1–40.

159. Rachel Carson, "Preface to Animal Machines [1964]," in *Lost Woods: The Discovered Writing of Rachel Carson*, ed. Linda Lear (Boston: Beacon Press, 1999), 193–194.

Chapter 4

1. David Biello, "Did the Anthropocene Begin in 1950 or 50,000 Years Ago?," *Scientific American*, April 2, 2015, https://www.scientificamerican.com/article/did-the-anthropocene-begin-in-1950-or-50-000-years-ago/.
2. Gerda Lerner, *The Creation of Patriarchy* (New York: Oxford University Press, 1986).
3. Susan Griffin, "Rape: The All-American Crime," *Ramparts Magazine*, Sept. 1971, 26–35, http://www.unz.org/Pub/Ramparts-1971sep-00026.
4. Jack D. Forbes, *Columbus and Other Cannibals: The Wétiko Disease of Exploitation, Imperialism and Terrorism*, Revised Edition (New York: Seven Stories Press, 1992), xvi. The original edition was published in 1979.
5. Martin Phillips, "What I Learned from One Mean Fucker," in *Male Lust: Pleasure, Power, and Transformation*, ed. Kerwin Kay, Jill Nagle, and Baruch Gould (New York: Harrington Park Press, 1999), 109–113, esp. 110.
6. Peggy Reeves Sanday, *Female Power and Male Dominance: On the Origins of Sexual Inequality* (Cambridge: Cambridge University Press, 1981). Sanday documents cultures that are free of rape; significantly, they are ones that also are the least destructive of ecosystems.
7. Rubin, *Deviations*, 33–65.
8. Lerner, Creation of Patriarchy, 212.
9. Lerner, *Creation of Patriarchy*, 213.
10. Jennifer A. Bennice and Patricia A. Resick, "Marital Rape: History, Research, and Practice," *Trauma, Violence, and Abuse* 4, no. 3 (2003): 228–246.
11. Anne Fausto-Sterling, "The Five Sexes, Revisited," *The Sciences*, July–August 2000, 19–23.

12. Andrea Dworkin, "The Rape Atrocity and the Boy Next Door," in *Our Blood: Prophecies and Discourses on Sexual Politics* (New York: Harper & Row, 1976), 22–49, esp. 45.

13. Lerner, *Creation of Patriarchy*, 9, 143.

14. Lerner, *Creation of Patriarchy*, 9.

15. Kim TallBear, "Making Love and Relations beyond Settler Sexualities," Social Justice Institute Noted Scholars Lecture Series, copresented by the Ecologies of Social Difference Research Network at the University of British Columbia, March 22, 2016, https://www.youtube.com/watch?v=zfdo2ujRUv8.

16. Leanne Betasamosake Simpson, *As We Have Always Done: Indigenous Freedom through Radical Resistance* (Minneapolis: University of Minnesota Press, 2017), 97.

17. Solomon Northrup, *12 Years a Slave*, ed. David Wilson (Chapel Hill: University of North Carolina Libraries, 2011); *12 Years a Slave*, dir. Steve McQueen (2013, Fox Searchlight Picture), DVD. Margaret Atwood, *The Handmaid's Tale* (Philadelphia: Chelsea House, 2011); *The Handmaid's Tale*, created by Bruce Miller (2017: Hulu), television series.

18. M. NourbeSe Philip, "Dis Place—The Space Between," in *A Genealogy of Resistance and Other Essays* (Toronto: Mercury Press, 1997), 74–112, esp. 92.

19. Jessica Bennett, Megan Twohey, and Alexandra Alter, "Why E. Jean Carroll, the 'Anti-Victim,' Spoke Up about Trump," *New York Times*, June 27, 2019.

20. Cynthia Enloe, *The Big Push: Exposing and Challenging the Persistence of Patriarchy* (Oakland: University of California Press, 2017).

21. Graham Readfern, "We Are Approaching the Trumpocene," *The Guardian*, Oct. 21, 2016.

22. David Moye and Daniel Marans, "Rashida Tlaib Only Sorry That Calling Trump a 'Motherf**ker' Has Been a 'Distraction,'" *Huffington Post*, Jan. 8, 2019, https://www.huffingtonpost.com/entry/rashida-tlaib-trump-m-fer_us_5c350c95e4b0dbd066012cf1.

23. Jane Caputi, "Sex and Violence in Popular Culture," in *A Companion to Popular Culture*, ed. Gary Burns (Malden, MA: Wiley-Blackwell, 2016), 421–440.

24. Richard Slotkin, *Regeneration through Violence: The Mythology of the American Frontier1600–1860* (Norman: University of Oklahoma Press, 2000, 1973). John Cawelti, "Violence and Apple Pie: Reflections on Literature, Culture and Violence," in *Mystery, Violence, and Popular Culture* (Madison: University of Wisconsin Popular Press, 2004), 210–216.

25. Smith, *Conquest*, 12.

26. André B. Rosay, "Violence against American Indian and Alaska Native Women and Men: 2010 Findings from the National Intimate Partner and Sexual Violence Survey," National Institute of Justice Research Report, May 2016, https://www.ncjrs.gov/pdffiles1/nij/249736.pdf.

27. Deer, *The Beginning and End of Rape*, 78.

28. Allen, *Sacred Hoop*, 36.

29. Sarah Deer and Liz Murphy, "'Animals May Take Pity on Us': Using Traditional Tribal Beliefs to Address Animal Abuse and Family Violence within Tribal Nations," *Mitchell Hamline Law Review* 43, no. 4 (2017): 702–742; Billy-Ray Belcourt, "Animal Bodies, Colonial Subjects: (Re)Locating Animality in Decolonial Thought," *Societies* 5, no. 1 (2015): 1–11, https://doi.org/10.3390/soc5010001.

30. Simpson, *As We Have Always Done*, 41.

31. Nick Estes, *Our History Is the Future: Standing Rock versus the Dakota Access Pipeline, and the Long Tradition of Indigenous Resistance* (New York: Verso, 2019).

32. Deer, *Beginning and End of Rape*, 117.

33. Steven Newcomb, "Perspectives: Healing, Restoration, and Rematriation," *News and Notes* (Spring–Summer 1995), 3. http://ili.nativeweb.org/perspect.html. See also *Rematriation Magazine*, https://www.facebook.com/Rematriation/.

34. Deb Haaland, "Trump Wants Immigrants to 'Go Back.' Native Americans Don't," *New York Times*, July 22, 2019.

35. Dianne D. Glave, *Rooted in the Earth: Reclaiming the African American Environmental Heritage* (Chicago: Lawrence Hill Press, 2010), 44.

36. Williams, "Sin, Nature, and Black Women's Bodies," 25, 29.

37. "Tar Sands," Indigenous Environmental Network, http://www.ienearth.org/what-we-do/tar-sands/. See also T. J. Demos,

Against the Anthropocene: Visual Culture and Environment Today (Berlin: Sternberg Press, 2017), 51–52.

38. Angela Y. Davis, *Women, Race and Class* (New York: Random House, 1981); Barbara Omolade, "Hearts of Darkness," in *Powers of Desire: The Politics of Sexuality*, ed. Ann Snitow, Christine Stansell, and Sharon Thompson (New York: Monthly Review Press, 1983), 350–367; Dorothy Roberts, *Killing the Black Body: Race, Reproduction, and the Meaning of Liberty* (New York: Pantheon Books, 1997). Adrienne Davis, "'Don't Let Nobody Bother Yo' Principle': The Sexual Economy of American Slavery," in *Sister Circle: Black Women and Work*, ed. Sharon Harley and the Black Women and Work Collective (Piscataway, NJ: Rutgers University Press, 2002), 103–127; Beth E. Richie, *Arrested Justice: Black Women, Violence, and America's Prison Nation* (New York: New York University Press, 2012); Threadcraft, *Intimate Justice*.

39. Thomas A. Foster, "The Sexual Abuse of Black Men under American Slavery," *Journal of the History of Sexuality* 20, no. 3 (2011): 445–464.

40. Threadcraft, *Intimate Justice*, 165, 156.

41. Davis, "'Don't Let Nobody Bother Yo' Principle,'" 105.

42. Davis, "'Don't Let Nobody Bother Yo' Principle,'" 105.

43. Frances S. Foster, "Ultimate Victims: Black Women in Slave Narratives," *Journal of American Culture* 1, no. 4 (1978): 845–854.

44. Janell Hobson, *Body as Evidence: Mediating Race, Globalizing Gender* (Albany: State University of New York Press, 2012), 142–145.

45. Darlene Hine and Kate Wittenstein, "Female Slave Resistance: The Economics of Sex," in *The Black Woman Cross-Culturally*, ed. Foomina C. Steady (Rochester, VT: Schenkman, 1981), 289–296.

46. Daina Ramey Berry, *The Price for Their Pound of Flesh: The Value of the Enslaved, from Womb to Grave, in the Building of a Nation* (Boston: Beacon Press, 2017).

47. Donna Haraway, Noboru Ishikawa, Scott F. Gilbert, Kenneth Olwig, Anna L. Tsing, and Nils Bubandt, "Anthropologists Are Talking—about the Anthropocene," *Ethnos* 81, no. 3 (2015): 535–564, doi:10.1080/00141844.2015.1105838. Patel and Moore, *History of the World*, 3.

48. On capitalism, see Patel and Moore, *History of the World*; Sven Beckert, "Slavery and Capitalism," *Chronicle of Higher Education*, Dec. 19, 2014, B4–B9.

49. Lee Maracle, *I Am Woman: A Native Perspective on Sociology and Feminism* (Vancouver: Press Gang, 1996, 139).

50. Linda Hogan, *Power* (New York: W. W. Norton, 1998), 69.

51. Alejandra Martinez, "Veteran Homelessness, Dolores Huerta and a True Crime Story," *Sundial*, WLRN-FM Public Radio, November 12, 2018, http://www.wlrn.org/post/veteran-homelessness-dolores-huerta-true-crime-story.

52. The Coalition of Immokolee Workers wages anti-slavery campaigns for farm workers and fights for fair pay. See "About CIW," Coalition of Immokolee Workers, http://ciw-online.org/about/.

53. Martin was a seventeen-year-old Black youth shot and killed by George Zimmerman, who was acquitted at trial. Brittney Cooper, *Eloquent Rage: A Black Feminist Discovers Her Superpower* (New York: St. Martin's Press, 2018), 183–184.

54. Lorraine Bayard deVolo and Lynn K. Hall, "'I Wish All the Ladies Were Holes in the Road': The US Air Force Academy and the Gendered Continuum of Violence," *Signs: Journal of Women in Culture and Society* 40, no. 4 (2015): 865–887.

55. Allison Fell, "One Night a Stranger . . . " *New Statesman and Society*, Oct. 24, 1991. Philip quotes this in "Dis Place," 76, 110n.

56. Jia Tolentino cites this from a poster on an incel message board: "Society has become a place for worship of females and it's so fucking wrong, they're not Gods they are just a fucking cumdumpster." "The Rage of the Incels," *The New Yorker*, May 15, 2018, https://www.newyorker.com/culture/cultural-comment/the-rage-of-the-incels.

57. Katherine McKittrick, *Demonic Grounds: Black Women and the Cartographies of Struggle* (Minneapolis: University of Minnesota Press, 2006), 44.

58. Garrard, *Brunelleschi's Egg*, 48.

59. Catharine A. MacKinnon, "Feminism, Marxism, Method, and the State: Toward Feminist Jurisprudence," *Signs: Journal of Women in Culture and Society* 8, no. 4 (Summer 1983): 635–658, esp. 636.

60. Catharine A. MacKinnon, "Desire and Power," in *Feminism Unmodified: Discourses on Life and Law* (Cambridge, MA: Harvard University Press, 1987), 46–62, esp. 55.

61. MacKinnon, "Feminism, Marxism, Method, and the State: Toward Feminist Jurisprudence," 636n.

62. Steve Russell, "The *Terra Nullius* Narrative Overturned," *Indian Country News*, June–July 2017, https://indian-countrymedianetwork.com/history/traditional-societies/terra-nullius-narrative-overturned.

63. Maureen Konkle, "Indigenous Ownership and Emergence of U.S. Liberal Imperialism," *American Indian Quarterly* 32, no. 3 (2008): 297–323, esp. 297.

64. Sharon Marcus, "Fighting Bodies, Fighting Words: A Theory and Politics of Rape Prevention," in *Theorizing Feminisms: A Reader*, ed. Elizabeth Hackett and Sally Haslanger (New York: Oxford University Press, 2006), 368–381, esp. 379.

65. Iako'tsi:rareh Amanda S. Lickers and Lindsay Nixon, "Terra Nullius Is Rape Culture," #LandBodyDefense, July 14, 2016, http://reclaimturtleisland.com/2016/07/14/terra-nullius-is-rape-culture-landbodydefense.

66. The *Urban Dictionary* gives one definition of the verb *drill*: "to root a chick hard," https://www.urbandictionary.com/define.php?term=Drill.

67. John Heidenry, *What Wild Ecstasy: The Rise and Fall of the Sexual Revolution* (New York: Simon and Schuster, 2002), 248.

68. I collected mostly anonymous commercial paraphernalia for the 2008 presidential election for a visual culture exhibit, *Three-Ring Political Circus*, Ritter Gallery, Florida Atlantic University, Sept. 5–Nov. 1, 2008. These items are now in the Visual Arts special collection in the Wimberly Library at Florida Atlantic University. One sticker depicts a silhouetted Palin working a spurting oil well as if it were a stripper pole, with the tag line "I'd drill that."

69. Timothy Cama, "Congress Votes to Open Alaska Refuge to Oil Drilling," *The Hill*, Dec. 20, 2017, http://thehill.com/policy/energy-environment/365772-congress-votes-to-open-alaska-refuge-to-oil-drilling.

70. A photo of this by Ruth Putter appeared on the back cover of the journal *Heresies: A Feminist Publication on Art and Politics* 5, no. 4 (1985). See also Caputi, *Gossips, Gorgons and Crones*, 43–44.

71. The *Oxford English Dictionary* provides this quote and its source: T. Killigrew, *Parsons Wedding* ii. vii, in *Comedies & Trag.* 107.

72. Posted by Senior Contributor, "The Court Jester," u/GayLubeOil, "How To Fuck Quasi Autistic Phone-Tarded Sluts In 2020," Red Pill Theory, Oct. 22, 2019, https://www.reddit.com/r/TheRedPill/comments/dlk5lb/how_to_fuck_quasi_autistic_phonetarded_sluts_in/.

73. Alberto Alesina, Paola Giuliano, and Nathan Nunn, "On the Origins of Gender Roles: Women and the Plough," *Quarterly Journal of Economics* 128, no. 2 (2013): 469–530. The theory connecting the plow to women's subordination was first put forth by Ester Boserup, *Woman's Role in Economic Development* (London: George Allen and Unwin, 1970).

74. Jackson Landers, "Did John Deere's Best Invention Spark a Revolution or an Environmental Disaster?," Smithsonian.Com, Dec. 17, 2015, https://www.smithsonianmag.com/smithsonian-institution/did-john-deeres-best-invention-spark-revolution-or-environmental-disaster-180957080/.

75. Katsi Cook, "Women Are the First Environment," *Indian Country Today*, Dec. 23, 2004, https://newsmaven.io/indian-countrytoday/archive/cook-women-are-the-first-environment-bZbKXN9CME-UabNOEhKgqg/.

76. Sjöö and Mor, *Great Cosmic Mother*, 384.

77. On plow culture, see Rosemary Radford Ruether, "Ecofeminism: Symbolic and Social Connections of the Oppression of Women and the Domination of Nature," in *This Sacred Earth: Religion, Nature, Environment,* 2nd ed., ed. Roger S. Gottlieb (New York: Routledge, 2004), 388–399, esp. 390.

78. Watts, "Indigenous Place-Thought," 21.

79. The website for the American Indian and Indigenous Studies Program at Cornell tells of a residential house dedicated to American Indian Studies, Akwe:kon. An unsigned essay explains: "Akwe:kon (A-gway-gohn) is a Mohawk word meaning "all of us". . . The arched forms of Akwe:kon suggest a similar motif found above the mail east entrance: the half-circle Sky Dome. The half-circle represents all life forms whom live in equality under the dome of the universe," https://aiisp.cornell.edu/akwekon/history-akwekon/symbols-and-meanings.

80. The quote is from an unsigned essay, "TERRA NULLIUS IS RAPE CULTURE," #LandBodyDefense, July 14, 2016, https:// reclaimturtleisland.com/2016/07/14/terra-nullius-is-rape-culture-landbodydefense/.

81. Watts, "Indigenous Place-Thought," 21.

82. Watts, "Indigenous Place-Thought," 21.

83. Watts, "Indigenous Place-Thought," 21.

84. Watts, "Indigenous Place-Thought," 20.

85. Kimmerer, *Braiding Sweetgrass*, 9.

86. "Feel the Raw Naked Power of the Road," Jensen car audio-speaker advertisement, jensenaudio.com, from about 1999.

87. Nick Estes pointed out the significance of settler mapping in a talk, "Native Resistance in the Age of Trump," Merrimack College, Massachusetts, March 14, 2018. See also Mishuana Goeman, *Mark My Words: Native Women Mapping Our Nations* (Minneapolis: University of Minnesota Press, 2013).

88. Jane Caputi. The Pornography of Everyday Life," in *Race, Class & Gender in the Media*, 4th edition, ed. Gail Dines and William E. Yousman (Sage Publications, 2018), 307–317. See also my short documentary, *The Pornography of Everyday Life*. http://www. berkeleymedia.com/product/the_pornography_of_everyday_ life/; Jane Caputi, "Is Seeing Believing: Rapist Screen Culture," in *Analyzing Violence against Women*, ed. Wanda Teays (Springer, 2019), 217–223.

89. "Highway Serial Killings: New Initiative on an Emerging Trend," FBI: Federal Bureau of Investigation, website, April 6, 2009, https://archives.fbi.gov/archives/news/stories/2009/april/high-wayserial_040609.

90. Deer, *Beginning and End of Rape*.

91. Ian Austen and Dan Bilefsky, "Canadian Inquiry Calls Killings of Indigenous Women Genocide," *New York Times*, June 3, 2019.

92. Terese Marie Mailhot, "When Will North America Reckon with the Ongoing Genocide of Indigenous Women," *Pacific Standard*, June 6, 2019, https://psmag.com/ideas/will-canada-reckon-with-the-ongoing-genocide-of-indigenous-women.

93. James Gilligan, *Violence: Reflections on a National Epidemic* (New York: Vintage Books, 1996), 267.

94. Tomson Highway, "Repairing the Circle: A Conversation with Tomson Highway," in *Masculindians: Conversations about Indigenous Manhood,* ed. Sam McKegney (Winnipeg, CAN: University of Manitoba Press, 2016), 21–29, esp. 27.

95. *Violence on the Land, Violence on Our Bodies,* 3.

96. *Violence on the Land, Violence on Our Bodies,* 62.

97. Katie Douglas, "Violence on our Earth," in *Violence on the Land, Violence on Our Bodies,* 5.

98. Katie Douglas discusses her work in "Everything Connected to the Land Is Connected to Our Bodies," WEA (Women's Earth Alliance," August 11, 2015, https://womensearthalliance.org/ wea-voices/everything-connnected-to-the-land-is-connected-to-our-bodies-2/.

99. MacKinnon, "The Art of the Impossible," in *Feminism Unmodified: Discourses on Life and Law* (Cambridge, MA: Harvard University Press, 1987), 1–17, esp. 6.

100. MacKinnon, "Art of the Impossible," 6–7.

101. Catharine A. MacKinnon, "Feminism, Marxism, Method, and the State: An Agency for Theory," *Signs: Journal of Women in Culture and Society* 7, no. 3 (1982): 515–544, esp. 541.

102. Swentzell makes this statement as part of a set of responses compiled by the Native American Studies Center, University of New Mexico, to the publication of *When Jesus came, the Corn Mothers Went Away: Marriage, Sex and Power in New Mexico, 1500–1846,* by Ramon A. Gutierrez. See "Commentaries," *American Indian Culture and Research Journal* 17, no. 3 (1991): 141–177, esp. 167.

103. Toni Morrison, *Beloved* (New York: Alfred A. Knopf, 1987), 87.

104. Michael Shellenberger and Ted Nordhaus, *Break Through: From the Death of Environmentalism to the Politics of Possibility* (New York: Houghton Mifflin, 2007), 7.

105. Shellenberger and Nordhaus, *Break Through,* 7.

106. For a discussion of "rapist culture," see Roxane Gay, "The Careless Language of Sexual Violence," in *Bad Feminist* (New York: Harper Perennial, 201), 128–136, esp. 133.

107. This is reported on the website of the Independence Institute, where Cooke is the Executive Vice President and Director of the Energy and Environmental Policy Center, https://i2i.org/ about/our-people/amy-oliver-cooke/.

108. "Fracking is Rape," Organelle.org, © 2011; "Frack Off," Climate Action Week: #FRACK OFF: Indigenous Women Leading Media Campaigns to Defend our Climate Action Week, September, 2014, http://events.newschool.edu/event/climate_action_week_frack_off_indigenous_women_leading_media_campaigns_to_defend_our_climate#; "Stop the Mother Frackers," can be viewed as one of the statements on a sign included in a number of anti-fracking images on Pinterest, https://www.pinterest.com/primevalelement/anti-fracking/.

109. Richard Manning, "Bakken Business: The Price of North Dakota's Fracking Boom," *Harper's*, March 2013, https://harpers.org/archive/2013/03/bakken-business/.

110. Awiakta, *Selu*, 16 and throughout.

111. Nocella, Bentley, and Duncan, "Introduction," xviii.

112. Teresa Brennan, *Exhausting Modernity: Grounds for a New Economy* (New York: Routledge, 2000), 41. Teresa Brennan, *The Transmission of Affect* (Ithaca, NY: Cornell University Press, 2004).

113. Jane Caputi, "Take Back What Doesn't Belong to Me": Sexual Violence and the 'Transmission of Affect,'" *Women's Studies International Forum* 26, no. 1 (2003): 1–14.

114. Nancy Venable Raine, *After Silence: Rape and My Journey Back* (New York: Three Rivers Press, 1998), 13.

115. Raine, *After Silence*, 225.

116. Patel and Moore, *History of the World*, 18–25.

117. Kim TallBear, *Native American DNA Tribal Belonging and the False Promise of Genetic Science* (Minneapolis: University of Minnesota Press, 2013).

118. Judith Bierman, untitled poem, in *She Who Was Lost Is Remembered: Healing from Incest through Creativity*, ed. Louise M. Wisechild (Seattle, WA: Seal Press, 1991), 195.

119. Marty O. Dyke, "Yeah, I'm Blaming You," in *I Never Told Anyone: Writings by Women Survivors of Child Sexual Abuse*, ed. Elaine Bass and Louise Thornton (New York: Harper Perennial, 1983), 113.

120. Jon Blistein, "Janice Dickinson on Bill Cosby: 'The Rape Is Etched into My Soul,'" *Rolling Stone*, Sept. 25, 2018, https://www.rollingstone.com/culture/culture-news/janice-dickinson-bill-cosby-rape-etched-my-soul-729321/.

121. Tarana Burke, "Inception," Just BE Inc., http://justbeinc.wixsite. com/justbeinc/the-me-too-movement-cmml.

122. Deer, *Beginning and End of Rape*, xvi.

123. *Violence on the Land, Violence on Our Bodies*, 63.

124. Amanda Kearney, *Violence in Place, Cultural and Environmental Wounding* (New York: Routledge, 2017), esp. 153–184.

125. *Violence on the Land, Violence on Our Bodies*, 63.

126. Nona Faustine, artist statement accompanying *White Shoes*, at the Smack Mellon Gallery, Brooklyn, New York, January 0 through February 21, 2016, http://www.smackmellon.org/exhi- bition/christine-sciulli-nona-faustine/#.

127. Jonathan Jones, "The Scars of America: Why a Nude Artist Is Taking a Stand at Slavery Sites," *The Guardian*, Aug. 5, 2015.

128. Linda Brent (Harriet Jacobs), *Incidents in the Life of a Slave Girl*, ed. L. Maria Child (New York: Harcourt Brace Jovanovich, 1973), 79.

129. Barbara Omolade, "Hearts of Darkness," in *Powers of Desire: The Politics of Sexuality*, ed. Ann Snitow, Christine Stansell, and Sharon Thompson (New York: Monthly Review Press, 1983), 350–367, esp. 351, 364.

130. Omolade, "Hearts of Darkness,"354.

131. Philip, "Dis Place," 96.

132. Philip, "Dis Place," 76.

133. Philip, "Dis Place," 110, note 4.

134. Philip, "Dis Place," 76.

135. Philip, "Dis Place," 77.

136. Ned Sublette and Constance Sublette, *The American Slave Coast: A History of the Slave-Breeding Industry* (Chicago: Lawrence Hill Books, 2016), xiii.

137. Sublette and Sublette, *American Slave Coast*, 24.

138. Sublette and Sublette, *American Slave Coast*, xiii.

139. Sublette and Sublette, *American Slave Coast*, xiv.

140. Farah Stockman, "Monticello Is Done Avoiding Jefferson's Relationship with Sally Hemings," *New York Times*, June 16, 2018.

141. Sublette and Sublette, *American Slave Coast*, 14.

142. Sublette and Sublette, *American Slave Coast*, 24.

143. Anna Tsing, "Earth Stalked by Man," *Cambridge Journal of Anthropology* 34, no. 1 (Spring 2016): 2–16, esp. 4, doi:10.3167/xa.2016.340102.

144. *Feminicide* is an intersectional elaboration (one also including state-supported violence) of the earlier concept of femicide. Rosa-Linda Fregoso and Cynthia Bejarano, "Introduction: A Cartography of Feminicide in the Américas," in *Terrorizing Women: Feminicide in the Américas*, ed. Rosa-Linda Fregoso and Cynthia Bejarano (Durham, NC: Duke University Press, 2010), 1–42.

145. Judith M. MacFarlane, Jacquelyn C. Campbell, Susan Wilt, Carolyn J. Sachs, Yvonne Ulrich, and Xiao Xu, "Stalking and Intimate Partner Femicide," *Homicide Studies* 3 no. 4 (1999): 300–316.

146. Edward E. Baptist, *The Half Has Never Been Told: Slavery and the Making of American Capitalism* (New York: Basic Books, 2014), 217.

147. Baptist, *Half Has Never Been Told*, 217.

148. Sven Beckert, "Slavery and Capitalism," *Chronicle of Higher Education*, Dec, 19, 2014, B4–B9.

Chapter 5

1. Tom Bee, "Color Nature Gone," lyrics by Tom Bee, used by permission, ©1973.

2. La Loba Loca, "Reclaiming Abuelita Knowledge as a Brown Ecofeminista," *Autostraddle*, March 20, 2014, https://www.autostraddle.com/reclaiming-abuelita-knowledge-as-a-brown-ecofeminista-213880/.

3. La Loba Loca, interview by Jane Caputi, August 6, 2015. Segments from this interview appear in my documentary *Feed the Green*. See also La Loba Loca, "Mamita Cuca: The Rape of the Sacred Coca Leaf," *BGD* (blog), July 8, 2013, https://www.bgdblog.org/2013/07/201378mamita-cuca-the-rape-of-the-sacred-coca-leaf/.

4. Cited in Cynthia Ozick, "How Helen Keller Learned to Write," *New Yorker*, June 8, 2003, https://www.newyorker.com/magazine/2003/06/16/what-helen-keller-saw.

5. Helen Keller and Annie Sullivan, *The Story of My Life* (New York: Grosset & Dunlap, 1905), 203.

6. John (Fire) Lame Deer and Richard Erdoes, *Lame Deer, Seeker of Visions* (New York: Simon and Schuster, 1972), 108–109.

7. Nigit'stil Norbert, "Beading Heart," *Edge YK*, June–July 2017, https://issuu.com/edgeyk/docs/edgeyk_junejuly_2017.

8. Chevalier and Gheerbrant, *Penguin Dictionary of Symbols*, 102.

9. Tito Naranjo and Rina Swentzell. "Healing Spaces in the Tewa Pueblo World," *American Indian Culture and Research Journal* 13, no. 3 (1989): 257–265, esp. 258.

10. "Cyanobacteria," June 21, 2012, *New World Encyclopedia*, http://www.newworldencyclopedia.org/entry/Cyanobacteria.

11. Martin Burd, "Colourful Language—It's How Aussie Birds and Flowers 'Speak,'" The Conversation, 2014, cited in Deborah Bird Rose and Thom van Dooren, "Encountering a More-Than-Human World: Ethos and the Arts of Witness," in *The Routledge Companion to the Environmental Humanities*, ed. Ursula K. Heise, Jon Christensen, and Michelle Nieman (New York: Routledge, 2017), 120–128.

12. Highway, "Repairing the Circle,' 29.

13. Omise'eke Natasha Tinsley, *Ezili's Mirrors: Imagining Black Queer Genders* (Durham, NC: Duke University Press, 2018), 9.

14. Kelly Grovier, "The History of the Rainbow Flag," *BBC*, June 15, 2016, http://www.bbc.com/culture/story/20160615-the-history-of-the-rainbow-flag.

15. Alex Abad-Santos, "Philadelphia's New, Inclusive Gay Pride Flag Is Making Gay White Men Angry," *Vox*, June 20, 2017, https://www.vox.com/culture/2017/6/20/15821858/gay-pride-flag-philadelphia-fight-explained.

16. Holland Carter, "'It's About Time!': Betye Saars's Long Climb to the Summit," *New York Times*, Sept. 4, 2019.

17. Allen, *Sacred Hoop*, 13–14.

18. Allen, *Sacred Hoop*, 2.

19. Paula Gunn Allen, *Grandmothers of the Light: A Medicine Woman's Sourcebook* (Boston: Beacon Press, 1991), 108.

20. Allen, *Sacred Hoop*, 14.

21. Paula Gunn Allen, "The Woman I Love Is a Planet, the Planet I Love Is a Tree," in *Reweaving the World: The Emergence of*

Ecofeminism, ed. Irene Diamond and Gloria Feman Orenstein (San Francisco: Sierra Club Books, 1990), 52–57, esp. 52.

22. Esther Leslie, *Synthetic Worlds: Nature, Art and the Chemical Industry* (Hammondsworth, UK: Reaktion Books), 47.

23. Toni Morrison, *The Bluest Eye* (New York: Alfred A. Knopf, 2000), 68, 92; Stallings, *Funk the Erotic*; Susan Willis, "Eruptions of Funk: Historicizing Toni Morrison," in *Black Literature and Literary Theory*, ed. Henry L. Gates Jr. (New York: Methuen, 1984), 263–283.

24. Morrison, *Bluest Eye*, 83.

25. Morrison, *Bluest Eye*, 131.

26. Cited in Derek Jarman, *Chroma: A Book of Color* (Woodstock, New York: The Overlook Press, 1994) 9.

27. Sylvia Wynter, "Sambos and Minstrels," *Social Text*, no. 1 (1979): 149–156.

28. Williams, "Sin, Nature, and Black Women's Bodies," 28.

29. Williams, "Sin, Nature, and Black Women's Bodies," 28.

30. Elijah Anderson, "The White Space," *Sociology of Race and Ethnicity* 1, no. 1 (2015): 10–21.

31. W. E. B. DuBois, *Darkwater: Voices from Within the Veil* (New York: Verso Books, 2016), 18.

32. Anne Spencer, "White Things," in *Black Nature: Four Centuries of African American Nature Poetry*, ed. Camille T. Dungy (Athens: University of Georgia Press, 2009), 155.

33. Morrison, *Beloved*, 89.

34. David Batchelor, *Chromophobia* (London: Reaktion Books, 2000).

35. See Shailja Patel, "Unpour," *Creative Time Reports*, May 8, 2014, http://creativetimereports.org/2014/05/08/shailja-patel-unpour-kara-walker-domino-sugar-factory/. A. Breeze Harper, "Social Justice Beliefs and Addiction to Uncompassionate Consumption: Food for Thought," in *Sistah Vegan: Black Female Vegans Speak Out on Food Identity, Health, and Society*, ed. A Breeze Harper (New York: Lantern Books, 2010), 20–41.

36. Cheryl L. Harris, "Whiteness as Property," *Harvard Law Review* 106, no. 8 (1993): 1710–1790.

37. I include images of such white spaces in my film *Feed the Green*.

38. *2001: A Space Odyssey*, dir. Stanley Kubrick (1968; California, Metro-Goldwyn-Mayer), film.

39. *Barbarella*, dir. Roger Vadim (1968; Paris, France: Marianne Productions), film.

40. For commentary and to view the ad, see Stephen Coles, "CompuServe Ad (1982): 'Welcome to Someday,'" *Fonts in Use*, April 7, 2013, https://fontsinuse.com/uses/3673/compuserve-ad-1982-welcome-to-someday.

41. Toni Morrison, *Tar Baby* (New York: Alfred A. Knopf, 1981), 203–204. I thank Helene Vann for alerting me to this passage.

42. Carl Zimring, *Clean and White: A History of Environmental Racism in the United States* (New York: New York University Press, 2015).

43. Alice Walker, "Everything Is a Human Being," in *Living by the Word: Selected Writings 1973–1987* (San Diego: Harcourt Brace Jovanovich, 1988), 139–152, esp. 147.

44. *Globopolis* sculpture by Elastic for *Wired*, Oct. 2015.

45. Ed Roberson, "We Must Be Careful," in *Black Nature: Four Centuries of African American Nature Poetry*, ed. Camille T. Dungy (Athens: University of Georgia Press, 2009), 3–5, esp. 5.

46. Rachel Carson, "The Real World around Us," in *Lost Woods: The Discovered Writing of Rachel Carson*, ed. Linda Lear (Boston: Beacon Press, 1999), 147–163, esp. 163.

47. Brennan, *Exhausting Modernity*. See Caputi, "On the Lap of Necessity: Myth and Technology through the Energetics Philosophy of Teresa Brennan," in *Goddesses and Monsters*, 289–314.

48. Fabiola López-Durán, *Eugenics in the Garden: Transatlantic Architecture and the Crafting of Modernity* (Austin: University of Texas Press, 2018), 43.

49. Greg Cajete, "An Enchanted Land: Spiritual Ecology and a Theology of Place," *Winds of Change* 8, no. 2 (Spring 1993): 50–55, esp. 51.

50. Luiz Edmundo, *O Rio de Janeiro do Meu Tempo*, 1938, cited in Sergio Burgi, "The Palace, the Square, and the Hill," *O Paço, a praça e o morro*, exhibit, Institute Moreira Salles, Rio de Janeiro, no date, https://artsandculture.google.com/exhibit/lgKyhIzkJQY8Jg.

51. López-Durán, *Eugenics in the Garden*, 59–60.

52. López-Durán, *Eugenics in the Garden*, 59–60.

53. López-Durán, *Eugenics in the Garden*, 58.

54. Alice Walker, "Coming Apart," in *Take Back the Night: Women on Pornography*, ed. Laura Lederer (New York: Bantam Books, 1980), 95–104.

55. J. Kohl, cartoon, *Hustler*, date unknown, ca. 1980.

56. Jennifer Nash, "Black Anality," *GLQ: Journal of Lesbian and Gay Studies* 20, no. 4 (2014): 439–460.

57. Josh Dawsey, "Trump Derides Protections for Immigrants from 'Shithole' Countries," *Washington Post*, Jan. 12, 2018, https://www.washingtonpost.com/politics/trump-attacks-protections-for-immigrants-from-shithole-countries-in-oval-office-meeting/2018/01/11/bfc0725c-f711-11e7-91af-35 31ac729add94_story.html?utm_term=.764cfa8200ca.

58. Fabiola López-Durán, *Eugenics in the Garden*, 64.

59. Fabiola López-Durán, *Eugenics in the Garden*, 64.

60. Jerry Dávila, *Diploma of Whiteness: Race and Social Policy in Brazil, 1917-1945* (Durham, NC: Duke University Press, 2003), 5.

61. The outrageous Valerie Solanas proffers this judgment: "The male, because of his obsession to compensate for not being female combined with his inability to relate and to feel compassion, has made of the world a shitpile." See *SCUM Manifesto*, in *I Shot Andy Warhol*, ed. Mary Harron and Daniel Minahan (New York: Grove Press, 1995), 161.

62. Suzanne Kelly, *Greening Death: Reclaiming Burial Practices and Restoring Our Tie to the Earth* (New York: Rowman & Littlefield, 2015), 41. See also Mel Y. Chen, *Animacies: Biopolitics, Racial Mattering, and Queer Affect* (Durham, NC: Duke University Press, 2012), 225.

63. Lev Grossman, "2145: The Year Man Becomes Immortal*," *Time*, Feb. 21, 2011. The cover photo-illustration is by Phillip Toledano for TIME, http://content.time.com/time/magazine/article/0,9171,2048299,00.html.

64. Alex Mar, "Love in the Time of Robots," *Wired*, Oct. 17, 2017, https://www.wired.com/2017/10/hiroshi-ishiguro-when-robots-act-just-like-humans/.

65. Robert W. Rydell, "Editor's Introduction 'Contend, Contend,'" in Ida B. Wells, Frederick Douglass, Irvine Garland Penn, and

Ferdinand L. Barnett, *The Reason Why the Colored American Is Not in the World's Columbian Exposition*, ed. Robert W. Rydell (Urbana: University of Illinois Press, 1999), xi–xlvii, esp. xi.

66. Erik Larson, *The Devil in the White City: Murder, Magic, and Madness at the Fair That Changed America* (New York: Vintage Books, 2003).

67. Quoted from the *Chicago Tribune* in Harold M. Mayer and Richard C. Wade, Chicago: *Growth of a Metropolis* (Chicago: University of Chicago Press. 1961), 196.

68. Robert W. Rydell, "The World's Columbian Exposition of 1893: Racist Underpinnings of a Utopian Artifact," *Journal of American Culture* 1, no. 3 (Summer 1978): 253–275, esp. 271.

69. Wells, Douglass, Penn, and Barnett, *Reason Why the Colored American Is Not in the World's Columbian Exposition*.

70. Rydell, "World's Columbian Exposition of 1893," 255.

71. McElya, *Clinging to Mammy*, 8–9.

72. Langston Hughes, "Lament for Dark Peoples," in *The Collected Poems of Langston Hughes*, ed. Arnold Rampersad (New York: Alfred A. Knopf, 1995), 39.

73. M. M. Manring, *Slave in a Box: The Strange Career of Aunt Jemima* (Charlottesville: University Press of Virginia, 1998), 75.

74. McElya, *Clinging to Mammy*, 3.

75. McElya, *Clinging to Mammy*, 207–252.

76. Alice Walker, "Giving the Party," *Anything We Love Can Be Saved: A Writer's Activism* (New York: Random House, 1997), 137–143, esp. 139.

77. Lucia Chiavola Birnbaum, *Dark Mother: African Origins and Godmothers* (San Jose, CA: Authors Choice Press), 2001; Charlotte Wood, "Finding Beauty in Broken Things, an Aboriginal Artist Finds Recognition at Last," *New York Times*, April 12, 2019.

78. Chevalier and Gheerbrant, *Penguin Dictionary of Symbols*, 163.

79. Walker, "Giving the Party," 141.

80. Walker, "Giving the Party," 143.

81. Carol Cohn, "Sex and Death in the Rational World of Defense Intellectuals," *Signs: Journal of Women in Culture and Society* 12, no. 4 (1987): 687–718.

82. "Collaborate in Bed," Bluebeam, ©2013 Bluebeam Software, Inc.

83. Henry Adams, *The Education of Henry Adams: A Centennial Version*, ed. Edward Chalfant and Conrad Edick Wright (Boston: Massachusetts Historical Society, 2007), 298.

84. Marina Warner, *Alone of All of Her Sex: The Myth and Cult of the Virgin Mary* (New York: Vintage Books, 1983), 280.

85. Simone de Beauvoir, *The Second Sex*, trans. and ed. H. M. Parshley (New York: Vintage Books, 1952), 193.

86. Daly, *Pure Lust*, 89–129.

87. Ean Begg, *The Cult of the Black Virgin* (New York: Penguin Books, 1985, 1996), 56.

88. Erich Neumann, *The Great Mother: An Analysis of the Archetype*, trans. Ralph Mannheim (Princeton, NJ: Princeton University Press, 1963), plate 161.

89. Warner, *Alone of Her Sex*, 274–280; Birnbaum, *Dark Mother*.

90. Adams, *Education of Henry Adams*, 301.

91. Kelly, *Greening Death*, 27–34.

92. Adams, *Education of Henry Adams*, 298.

93. McLuhan, *Mechanical Bride*, 101.

94. Jenny Kleeman, "The Race to Build the World's First Sex Robot," *The Guardian*, April 27, 2017.

95. Sarah Sharma, "Going to Work in Mommy's Basement," *Boston Review Forum* 7 June 19, 2018, http://bostonreview.net/gender-sexuality/sarah-sharma-going-work-mommys-basement.

96. Sharma, "Going to Work," 91.

97. The word *fembot*, meaning a female robot, first appeared in an episode of the television show *The Bionic Woman* (1976). For a feminist definition, see Daly with Caputi, *Wickedary*, 198.

98. The photo-illustration for the April 15 cover of *Kiplinger's* is by Colin Anderson. See Kiplinger's Personal Finance, Kiplinger website, https://www.kiplinger.com/fronts/archive/magazine/index.html?month=04&year=2015.

99. Ntozake Shange, "We Need a God Who Bleeds Now," in *A Daughter's Geography* (New York: St. Martin's Press, 1983), 51.

100. Gregory Jerome Hampton, *Imagining Slaves and Robots in Literature, Film, and Popular Culture: Reinventing Yesterday's Slave with Tomorrow's Robot* (Lanham: Lexington Books, 2015).

101. "Meet Bixby," Samsung, https://www.samsung.com/us/explore/bixby/.

102. Thanks to Toni Francis for this observation.

103. Hampton, *Imagining Slaves and Robots*, x.

104. McKittrick, *Demonic Grounds*, 46.

105. David Levy, *Robots Unlimited: Life in a Virtual Age* (Wellesley, MA: A. K. Peters, 2006), 343.

106. Henry T. Greely, *The End of Sex and the Future of Human Reproduction* (Cambridge, MA: Harvard University Press, 2016), 188.

107. No Limits™" is the slogan for Bluebeam Software, "The Nature of What's to Come" was used by ADM in the early 21st century. See Emily Wilson, "ADM launches new corporate logo, campaign," World-Grain.com, May 1, 2001, http://www.world-grain.com/News/Archive/ADM-launches-new-corporate-logo-campaign.aspx?cck=1. "The Miracles of Science™" belongs to DuPont; "We Bring Good Things to Life" is the old slogan of General Electric; "Where the Future Takes Shape" identifies Arcona, https://www.arconic.com/global/en/home.asp; Coca-Cola uses "It's the Real Thing; and Lockheed Martin uses "Engineering a Better Tomorrow," as its trademark, https://www.lockheedmartin.com/en-us/index.html.

108. Caoimhe McL, "Plastic Soup—a Lethal Blend" (blog post), Jan. 1, 2016, http://my5pence.blogspot.com/2016/01/plasticocene.html.

109. Patricia L. Corcoran, Charles J. Moore, and Kelly Jazvac, "An Anthropogenic Marker Horizon in the Future Rock Record," *GSA Today* 24, no. 6 (June 2014), 4–8. |

110. Jennifer Gabrys, Gay Hawkings, and Mike Michael, "Introduction: From Materiality to Plasticity," in *Accumulation: The Material Politics of Plastic*, ed. Jennifer Gabrys, Gay Hawkins, and Mike Michael (New York: Routledge, 2013), 1–14, esp. 11.

111. Maureen E. Butter, "Are Women More Vulnerable to Environmental Pollution?," *Journal of Human Ecology* 20, no. 3 (2006): 221–226.

112. E. B. Couzens and V. E. Yarsley, *Plastics in the Modern World: A Completely Revised Edition of Plastics in the Service of Man* (Baltimore: Penguin Books, 1968), 31.

113. Leslie, *Synthetic Worlds*, 7.

114. Heather Davis, "Life and Death in the Anthropocene: A Short History of Plastic," in *Art in the Anthropocene: Encounters among Aesthetics, Politics, Environment and Epistemology*, ed. Heather Davis and Etienne Turpin (London: Open Humanities Press, 2015), 347–358.

115. Max Liboiron, "Plasticizers: A Twenty-First Century Miasma," in *Accumulation: The Material Politics of Plastic*, ed. Jennifer Gabrys, Gay Hawkins, and Mike Michael (New York: Routledge, 2013), 134–149, esp. 139.

116. Leslie, *Synthetic Worlds*, 47.

117. Royce, "Usage of Chemicals in Chemical Industry," June 28, 2013, https://www.slideshare.net/Royceintly/plastic-colorant-and-its-role-in-plastic-industry.

118. Leslie, *Synthetic Worlds*, 246.

119. Leslie, *Synthetic Worlds*, 47.

120. V. E. Yarsley and E. G. Couzens, "Plastics and the Future," in *The Plastics Age: From Bakelite to Beanbags and Beyond*, ed. Penny Sparke (Woodstock, NY: Overlook Press, 1992), 55–59, esp. 55, 58.

121. Fortune, "Plastics in 1940," *Fortune*, Oct. 1940, 92–93, 88–89.

122. "Synthetica—a New Continent of Plastics." *Fortune*, 1940. Executed by Ortho Plastics Novelties, Inc, in *Fortune* 22 (October 1940), 92–93.

123. Department of the Environment, Water, Heritage and the Arts, "Formaldehyde," National Pollutant Inventory, Australian Government Department of Energy and the Environment, Canberra, AUS, 2009, http://www.npi.gov.au/resource/formaldehyde.

124. Jeffrey L. Meikle, *American Plastic: A Cultural History* (New Brunswick, NJ: Rutgers University Press, 1995), 66.

125. Merchant, *Earthcare*, 84.

126. Caputi, *Age of Sex Crime*, 1987.

127. Tom Cullen, *When London Walked in Terror* (Boston: Houghton Mifflin Company, 1965), 14.

128. Jane Caputi, "Sex and Violence in Popular Culture, in *A Companion to Popular Culture*, ed. Gary Burns (Malden, MA: Wiley-Blackwell, 2016), 421–440.

129. *Dr. Strangelove, or: How I Learned to Stop Worrying and Love the Bomb*, dir. Stanley Kubrick (1964. Columbia Pictures), film.

130. "An American Dream of Venus," *Fortune*, 1940. Photo for *Fortune* by Matter-Bourges. Copyright 1940 by Time, Inc. All Rights Reserved, 88–89.

131. Diane Wolkstein and Samuel Noah Kramer, *Inanna: Queen of Heaven and Earth: Her Stories and Hymns from Sumer* (New York: Harper Perennial, 1983), 169.

132. Garrard, *Brunelleschi's Egg*, 89.

133. Sandro Botticelli, "The Birth of Venus," 1484-1486, Uffizi Gallery, Florence, Italy.

134. David Kingsley, *The Goddesses' Mirror* (Albany: State University of New York Press, 1989), 189.

135. Randy P. Conner, David Hatfield Sparks, and Mariya Sparks, *Cassell's Encyclopedia of Queer Myth, Symbol, and Spirit: Gay, lesbian, bisexual, and transgender lore* (London: Cassell, 1997). 258–259, 341–342.

136. Saidiya Hartman, "Venus in Two Acts," *Small Axe* 12 (2):1–14.

137. Janell Hobson, *Venus in the Dark: Blackness and Beauty in Popular Culture*, 2nd ed. (New York: Routledge, 2018).

138. Syreeta McFadden, "The Lack of Female Genitals on Statues Seems Thoughtless until You See it Repeated," *The Guardian*, April 13, 2015.

139. Meikle, *American Plastic*, 66–67.

140. Meikle, *American Plastic*, 67.

141. Tim Murphy, "Gay Men and Misogyny: Rose McGowan's Half-Right," *The Cut*, Nov. 9, 2013, https://www.thecut.com/2014/11/gay-men-and-misogyny-rose-mcgowans-half-right.html.

142. Aurora Levins Morales, *Kindlings: Writings on the Body* (Cambridge, MA: Palabera Press, 2013), 61–76. Mel Y. Chen, *Animacies: Biopolitics, Racial Mattering, and Queer Affect* (Durham, NC: Duke University Press, 2012).

143. See Murphy, "Gay Men and Misogyny." See also *The Online Slang Dictionary* definition of *fish* as "a drag queen that looks so much like a woman you can almost smell a fishy vagina," http://onlineslangdictionary.com/meaning-definition-of/fish.

144. Meikle, *American Plastic*, 64.

145. Meikle, *American Plastic*, 67.

146. Sigmund Freud, "Dreams," in *The Standard Edition of the Complete Psychological Works of Sigmund Freud*, vol. 15 (London: Hogarth, 1953), 154.

147. Highway, "Repairing the Circle," 27.

148. Edwin Slosson, quoted in Meikle, *American Plastic*, 69.

149. McLuhan, *Mechanical Bride*, 101.

150. Chevalier and Gheerbrant, *Penguin Dictionary of Symbols*, 783.

151. Ben Lovejoy, "PSA: Apple's Renewal of Its Trademark for the Rainbow Apple Logo Doesn't Mean Anything," *9TO5Mac*, Feb. 20, 2018, https://9to5mac.com/2018/02/20/apple-rainbow-apple-logo/.

152. *Wired*, December 2015. Cover photograph by Richard Mosse, https://www.wired.com/story/wired-cover-browser-2015/.

153. J. Weston Phippen, "The Navajo, the EPA, and the Accident That Turned a River Orange," *The Atlantic*, Aug, 16, 2016, https://www.theatlantic.com/news/archive/2016/08/navajo-nation-lawsuit-gold-king-mine-spill/496073/.

154. Kim Stanley Robinson, *Blue Mars* (London: HarperCollins, 1996), 672–673.

155. Bronislaw Szerszynski, "Coloring Climates: Imagining a Geoengineered World," in *The Routledge Companion to the Environmental Humanities*, ed. Ursula K. Heise, Jon Christensen, and Michelle Nieman (New York: Routledge, 2017), 82–90.

156. Franklin Foer, *World without Mind: The Existential Threat of Big Tech* (New York: Penguin Press, 2017), 12.

157. Nocella, Bentley, and Duncan, "Introduction," xix.

158. Jody A. Roberts, "Reflections of an Unrepentant Plastiphobe: An Essay on Plasticity and the STS Life," in *Accumulation: The Material Politics of Plastic*, ed. Jennifer Gabrys, Gay Hawkins, and Mike Michael (New York: Routledge, 2013), 121–133.

159. Roberts, "Reflections of an Unrepentant Plastiphobe," 130.

160. Walker, "Everyone Is a Human Being," 148.

161. Sherry Turkle, *Reclaiming Conversation: The Power of Talk in a Digital Age* (New York: Random House, 2015).

162. Kyle Powys Whyte, Ryan Gunderson, and Brett Clark, "Is Technology Insidious?," in *Philosophy, Technology, and the Environment*, ed. David M. Kaplan (Cambridge, MA: MIT Press, 2017), 41–61, esp. 41.

163. Whyte, Gunderson, and Clark, "Is Technology Insidious?," 41.
164. Aldo Leopold, "Good Oak," *A Sand County Almanac: With Essays on Conservation from Round River* (New York: Ballantine Books, 1949), 6.
165. Roland Barthes, "Plastic," in *Mythologies*, trans. Annette Lavers (New York: Hill and Wang, 1972), 97–90.
166. Pamela Gupta and Gabrielle Hecht, "Toxicity, Waste, Detritus: An Introduction," *Somatosphere*, Oct. 10, 2017, http://somatosphere.net/2017/10/toxicity-waste-detritus-an-introcduction.html.
167. The dream was on February 18, 2013.

Chapter 6

1. Awiakta, "When Earth Becomes an 'It,'" in *Selu*, 6.
2. Chevalier and Gheerbrant, *Penguin Dictionary of Symbols*, 451.
3. Jonathan Watts, "Scientists Identify Vast Underground Ecosystem Containing Billions of Micro-organisms," *The Guardian*, Dec. 10, 2018.
4. Oliver Morton, *Eating the Sun: How Plants Power the Planet* (New York: HarperCollins, 2008).
5. Morton, *Eating the Sun*, xv.
6. Carl Zimmer, "Cyanobacteria Are Far from Just Toledo's Problem," *New York Times*, Aug. 7, 2014.
7. Simon Ortiz, "We Have Been Told Many Things but We Know This to Be True," in *Woven Stone* (Tucson: University of Arizona Press, 1992), 324–325.
8. N. Scott Momaday, "Native American Attitudes to the Environment," in *Seeing with a Native Eye: Essays on Native American Religion*, ed. W. H. Capps (New York: Harper & Row, 1976), 79–85, esp. 83.
9. Wangari Maathai, *Replenishing the Earth: Spiritual Values for Healing Ourselves and the World* (New York: Doubleday, 2010).
10. Winona LaDuke, "A Society Based on Conquest Cannot Be Sustained," in *The New Resource Wars: Native and Environmental Struggles against Multinational Corporations*, ed. Al Gedicks (Boston: South End Press 1993, ix–xv, esp. x.

11. Patel and Moore, *History of the World in Seven Cheap Things*, 19–20.
12. Smitherman, *Black Talk*, Revised Edition, 126.
13. Dave Walker, "Witness Recalls Role of New Orleans' Mahalia Jackson in Martin Luther King's 'I Have a Dream' Speech," *New Orleans Times-Picayune*, Aug. 23, 2013.
14. Naomi Klein, "Dancing the World into Being: A Conversation with Idle No More's Leanne Simpson," *Yes Magazine*, March 5, 2013, https://www.yesmagazine.org/peace-justice/dancing-the-world-into-being-a-conversation-with-idle-no-more-leanne-simpson.
15. Numerous lists of green pop figures are available on the Internet. One is compiled by the *Los Angeles Times*, "Top 15 Green Characters in Pop Culture," n.d., http://www.latimes.com/entertainment/la-green-characters-pg-photogallery.html.
16. Chevalier and Gheerbrant, *Penguin Dictionary of Symbols*, 454–455. The Holy Grail is spoken of as a green vessel filled with red blood.
17. This button can be viewed at https://www.etsy.com/no-en/listing/274863796/mother-nature-is-a-lesbian-15-button. Another version of Cowan's button is available for sale from Small Equals, her shop in Burlington, Vermont, and online at small-equals.com, https://www.smallequals.com/listing/274863796/mother-nature-is-a-lesbian-feminist, and through her Small Equals shop on Etsy.
18. *Moana*, directed by Ron Clements & John Musker (2016; Burbank, CA: Walt Disney Animation Studios) film. Doug Herman, "How the Story of 'Moana' and Maui Holds Up against Cultural Truths," *Smithsonian.com*, December 2, 2016, https://www.smithsonianmag.com/smithsonian-institution/how-story-moana-and-maui-holds-against-cultural-truths-180961258/.
19. Baring and Cashford, *Myth of the Goddess*, 176.
20. Baring and Cashford, *Myth of the Goddess*, 364.
21. Chevalier and Gheerbrant, *Penguin Dictionary of Symbols*, 454.
22. Richard H. Wilkinson, *Symbol and Magic in Egyptian Art* (New York: Thames and Hudson, 1994), 108.
23. Baring and Cashford, *Myth of the Goddess*, 237.

24. Nawal el Saadawi, *A Daughter of Isis* (London: Zed Books, 1999), 14.
25. Chevalier and Gheerbrant, *Penguin Dictionary of Symbols*, 452–453.
26. Miranda Shaw, *Buddhist Goddesses of India* (Princeton, NJ: Princeton University Press, 2006), 324.
27. From the "108 Praises of Tārā," cited in Shaw, *Buddhist Goddesses of India*, 306.
28. Shaw, *Buddhist Goddesses of India*, 326.
29. Paul Gauguin, *The Green Christ*, 1889, oil paint, 3' 0" x 2' 5", Royal Museums of Fine Arts of Belgium.
30. Katharine Briggs, *The Encyclopedia of Fairies, Hobgoblins, Brownies, Bogies, and Other Supernatural Creatures* (New York: Pantheon, 1976), 108.
31. John Hutchings, "Folklore and Symbolism of Green," *Folklore* 108 (1997): 55–64, esp. 59
32. Hutchings, "Folklore," 59
33. Malidoma Patrice Somé, *Of Water and the Spirit: Ritual, Magic, and Initiation in the Life of an African Shaman* (New York: G. P. Putnam's Sons, 1994), 202–203.
34. Alexis Brooks De Vita, *Mythatypes: Signatures and Signs of African/Diaspora and Black Goddesses* (Westport, CT: Greenwood Press, 2000), 33.
35. Cited in Susan Cady, Marian Ronan, and Hal Taussig, *Sophia: The Future of Feminist Spirituality* (San Francisco: Harper & Row, 1986), 28.
36. Andrew Harvey and Anne Baring, *The Divine Feminine: Exploring the Feminine Face of God throughout the World* (Berkeley, CA: Conari Press, 1996), 100.
37. Cady, Ronan, and Taussig, *Sophia*, 29
38. This is quoted by Edwin M. Loeb as told to him by an "East Pomo informant Benson." See Edwin M. Loeb, *Pomo Folkways* (Berkeley: University of California Press, 1926), 302.
39. Shel Silverstein, *The Giving Tree* (New York: Harper Collins, 2004, 1964).
40. *Mother!*, dir. Darren Aronofsky (2017; New York, NY; Protozoa Pictures) film.

41. Superman's nemesis kryptonite, Kermit the Frog, Yoda, Gumby, Godzilla, Dr. Doom, the Mask, the Grinch, Gamora, Slimer, Oscar the Grouch, Ghoulies, Green Arrow, the Green Lantern, the Jolly Green Giant, the Green Goblin, Monster Energy Drink, and beloved characters like Shrek and the red-haired Fiona. For comic relief, there is the raucous mucous character in Mucinex ads,

42. Gregory Maguire, *Wicked: The Life and Times of the Wicked Witch of the West* (New York: Regan Books, 1995).

43. *Sleeping Beauty,* dir. Clyde Geronimi (1959, Walt Disney Productions) DVD.

44. Poster for a play of Bram Stoker's story, held by Heritage Auction Galleries. See Joan Acocella, "In the Blood: Why Vampires Still Thrill," *New Yorker,* March 16, 2009, https://www.newyorker.com/magazine/2009/03/16/in-the-blood.

45. Georg Pencz, illustration for Hans Sachs, Das Feindtselig Laster, dey heymlich Neyd (Nuremberg, 1534), in Lyndal Roper, *The Witch in the Western Imagination* (Charlottesville: University of Virginia Press, 2012), 95.

46. Jarman, *Chroma,* 63.

47. Gloria Feman Orenstein, "Recovering Her Story: Feminist Artists Reclaim the Great Goddess," in *The Power of Feminist Art,* ed. Norma Broude and Mary D. Garrard (New York: Harry N. Abrams, 1994), 174–189, esp. 187.

48. Alison Saar, *Mamba Mambo,* 1985, mixed media, 5' 4" x 2' 10" x 1' 5." Reproduced in Lisa E. Farrington, *Creating Their Own Image: The History of African-American Women Artists* (New York: Oxford University Press, 2005), 263.

49. Farrington, *Creating Their Own Image,* 263.

50. Farrington, *Creating Their Own Image,* 264.

51. Morton, *Eating the* Sun, xv.

52. Lao Tzu, *The Complete Works of Lao Tzu: Tao Teh Ching and Hua Hu Ching,* trans. Hua Ching Ni (Los Angeles: Sevenstar Communications, 1995), 12. First edition published in 1979.

53. *The Fantastic Four,* dir. Josh Trank (2015; Los Angeles, CA: Twentieth Century Fox), film.

54. Aldo Leopold, "The Land Ethic," in *A Sand County Almanac: With Essays on Conservation from Round River* (New York: Ballantine Books, 1966), 253.

55. David Greenberg, *Magic* (Niles, IL: DC Comics, Mayfair Games Inc., 1992), 125

56. Martin Buber, *I and Thou*, trans. Walter Kaufmann (New York: Touchstone, 1996).

57. Gilligan, *Violence*, 36 and throughout.

58. Patel and Moore, *History of the World in Seven Cheap Things*, 19–20.

59. Morrison, *Beloved*, 28.

60. Morrison, *Beloved*, 89.

61. Morrison, *Beloved*, 28–29.

62. Dianne D. Glave, *Rooted in the Earth: Reclaiming the African American Environmental Heritage* (Chicago: Lawrence Hill Press, 2010), 44.

63. Glave, *Rooted in the Earth*, 56.

64. Starhawk, "Ritual as Bonding: Action as Ritual," in *Weaving the Visions: New Patterns in Feminist Spirituality*, ed. Judith Plaskow and Carol P. Christ (New York: Harper and Row, 1989), 326–335.

65. Starhawk, *Webs of Power*, 160–167.

66. Carol Lee Sanchez, "New World Tribal Communities: An Alternative Approach for Recreating Egalitarian Societies," in *Weaving the Visions*, 344–356, esp. 345–346.

67. Kimmerer, *Braiding Sweetgrass*, 9.

68. Klein, "Dancing the World into Being."

69. Jane Caputi, "Feeding Green Fire," *Journal for the Study of Religion, Nature, and Culture* 5, no. 4 (2011): 410–436.

70. Chevalier and Gheerbrant, *Penguin Dictionary of Symbols*, 455.

71. Anne H. King-Lenzmeier, *Hildegard of Bingen: An Integrated Vision* (Collegeville, MN: Liturgical Press, 2001), 6.

72. Patricia Monaghan, *The Goddess Companion: Daily Meditations on the Feminine Spirit* (St. Paul, MN: Llewellyn, 1999), 123.

73. Barbara Newman, *Hildegard of Bingen Symphonia*, 2nd ed. (Ithaca, NY: Cornell University Press), 199.

74. Tanya Torres, *Atabey Gives Birth to the Coqui*, Oil on Canvas, 20" x 16," 2012, http://tanyatorres.blogspot.com/2012/04/atabey-gives-birth-to-coqui.html.

75. Edouard Duval-Carrié, *Primitif Futur*, 2002, in *Duval-Carrié: La Casa En Llama*, Lyle O. Reitzel Arte Contemporaneo, 2004.

76. Bron Taylor, conversation with the author, Oshkosh, Wisconsin, March 28, 2001. Bron Taylor, *Dark Green Religion* (Berkeley: University of California Press, 2010).

77. Dan Shilling, "Aldo Leopold Listens to the Southwest," *Journal of the Southwest* 51, no 2 (2009): 317–350.

78. Aldo Leopold, "Thinking Like a Mountain," in *Sand County Almanac*, 137–141, esp. 138–39.

79. Jody Emel, "Are You Man Enough, Big and Bad Enough? Wolf Eradication in the US," in *Animal Geographies: Place, Politics, and Identity in the Nature-Culture Borderlands*, ed. Jennifer Woldch and Jody Emel (New York: Verso, 1998), 91–116.

80. Leopold, "Thinking Like a Mountain," 140.

81. Juan Luis Arsuaga and Ignacio Martínez, *Green Fire: The Life Force, from the Atom to the Mind*, trans. Michael B. Miller (New York: Thunder's Mouth Press, 2004).

82. Email correspondence, Feb. 18, 2010. Thanks to Michael Miller, the translator, for directing me to Mr. Oakes.

83. Dylan Thomas, "The Force That through the Green Fuse Drives the Flower," in *The Collected Poems of Dylan Thomas* (New York: New Directions, 1957), 10–11.

84. Kingsley, *Goddesses' Mirror*, 189.

85. Kimmerer, *Braiding Sweetgrass*, 48–49.

86. Kimmerer, *Braiding Sweetgrass*, 49.

87. Levingston, Steven. "Jellyfish Gene Leaps from the Test Tube to Tickle Entrepreneurs' Imagination," *New York Times*, Jan. 18, 2001.

88. Eduardo Kac, "GFP Bunny," http://www.ekac.org/gfpbunny.html#gfpbunnyanchor, published in Peter T. Dobrila and Aleksandra Kostic, Aleksandra, eds., *Eduardo Kac: Telepresence, Biotelematics, and Transgenic Art* (Maribor, Slovenia: Kibla, 2000), 101–131.

89. Sheilah Britton and Dan Collins, eds., *The Eighth Day: The Transgenic Art of Eduardo Kac* (Tempe, AZ: Institute for Studies in the Arts, 2003).

90. Significantly, Kac's GFP Bunny seems to be modeled after Andy Warhol's *Chocolate Bunny*, an image of a green-foil wrapped Easter bunny. Andy Warhol, *Chocolate Bunny*, 1983, synthetic polymer and silkscreen inks on canvas, 20 x 16 in.

91. Chevalier and Gheerbrant, *Penguin Dictionary of Symbols*, 472.

92. Jeffrey J. Cohen, "Afterword: An Unfinished Conversation about Glowing Green Bunnies," *Queering the Non/Human*, ed. Noreen Giffney and Myra J. Hird (Burlington. VT: Ashgate, 2008), 363–376.

93. For pertinent thoughts on integrity, see Daly, *Quintessence*, 11 and throughout; Threadcraft, *Intimate Justice*, 156 and throughout.

94. Kac, *Telepresence*, 271.

95. Douglas Martin, "Margaret Tafoya, 96, Pueblo Potter Whose Work Found a Global Audience," *New York Times*, March 5 2001.

96. Ted Chiang, "Story of Your Life," in *Arrival* (New York: Vintage Books, 2016, 2002), 91–146, esp. 114.

97. United Nations, 2019 International Year of Indigenous Languages, https://en.iyil2019.org/.

98. Kimmerer, *Braiding Sweetgrass*, 55.

99. Robin Wall Kimmerer, "Nature Needs a New Pronoun: To Stop the Age of Extinction, Let's Start by Ditching 'It,'" *Yes Magazine*, March 30, 2015, http://www.yesmagazine.org/issues/together-with-earth/alternative-grammar-a-new-language-of-kinship.

100. Kimmerer, *Braiding Sweetgrass*, 55.

101. Kimmerer, *Braiding Sweetgrass*, 51.

102. Kimmerer, "Nature Needs a New Pronoun."

103. Ellie Shechet, "Do Plants Have Something to Say?," *New York Times*, Aug. 29, 2019.

104. Kimmerer, *Braiding Sweetgrass*, 59.

105. Norbert, "Beading Heart," 9.

106. Evelyn Fox Keller, "A World of Difference," in *Reflections on Gender and Science* (New Haven, CT: Yale University Press, 1985), 158–176, esp.162.

107. Keller, "World of Difference," 164.

108. Naomi Oreskes, "Put Government Labs to Work on Climate Change," *Washington Post*, Jan. 17, 2013.

109. Starhawk, filmed interview with Jane Caputi, San Francisco, June 17, 2013.

110. Kimmerer, *Gathering Moss*, 159.

111. Kimmerer, *Gathering Moss*, 158.

112. Kimmerer, *Gathering Moss*, 161.

113. Kimmerer, *Gathering Moss*, 162.

114. Leopold, *Sand County Almanac*, 254.

115. Helen Czerski is a physicist who was on board the Swedish ice-breaker Oden in the summer of 2018 and was interviewed on a program broadcast by NPR (WLRN), which I heard but was unable to get any specifics.

116. W. Steffen, A. Sanderson, P. D. Tyson, J. Jäger, P. A. Matson, B. Moore III, F. Oldfield, K. Richardson, H. J. Schellnhuber, B. L. Turner, R. J. Wasson, "Executive Summary," *Global Change and the Earth System: A Planet Under Pressure*, 2004, 19, http://www.igbp.net/download/18.1b8ae20512db692f2a680007761/1376383137895/IGBP_ExecSummary_eng.pdf. This report was sponsored by the International Geosphere-Biosphere Programme. The executive summary is available online. The full book is published by Springer-Verlag Berlin Heidelberg New York, 2005.

117. Forbes, "Nature and Culture," 107.

118. Lee Maracle, "Decolonizing Native Women," in *Daughters of Mother Earth: The Wisdom of Native American Women*, ed. Barbara Alice Mann (Westport, CT: Praeger, 2006), 29–51, esp. 43.

119. Little Bear, "Foreword," x–xi.

120. Illustration by Tracy Ma / *The New York Times*; Ford Fischer. The illustration accompanies Jonah Engel Bromwick, "Death of a Biohacker," *New York Times*, May 19, 2018.

121. Mary Shelley, *Frankenstein; Or, The Modern Prometheus* (New York: Pocket Books, 2004).

122. Jonathan Bailey, "How Universal Re-copyrighted Frankenstein's Monster," *PT Plagiarism Today*, Oct. 24, 2011, https://www.plagiarismtoday.com/2011/10/24/how-universal-re-copyrighted-frankensteins-monster/.

123. Dr. Seuss, *Benjamin and the Oobleck* (New York: Random House, 1949), n.p..

124. Jim Cheney and Anthony Weston, "Environmental Ethics as Environmental Etiquette: Towards an Ethics-Based Epistemology," *Environmental Ethics* 21 (1999): 115–134.

125. Lewis Hyde, *The Gift: Imagination and the Erotic Life of Property* (New York: Random House, 1983).

126. Jeff Vandermeer, *Area X: The Southern Reach Trilogy* (New York: Farrar, Straus and Giroux, 2014), 92. *The Annihilation*, dir. Alex Garland (2018; Los Angeles, CA: Paramount Pictures) film.

127. Vandermeer, *Area X*, 92.

128. Louise Erdrich, *Future Home of the Living God* (New York: HarperCollins, 2017), 3.

129. Erdrich, *Future Home*, 13.

130. Haraway, *Staying with the Trouble*, 140.

131. Haraway, *Staying with the Trouble*, 139.

132. Haraway, *Staying with the Trouble*, 141

133. Haraway, *Staying with the Trouble*, 140.

134. Kara Marker, "The CRISPR Craze Paints the Future of Genome Engineering," *Labroots*, Oct. 12, 2016, https://www.labroots.com/trending/health-and-medicine/4302/crispr-craze-paints-future-genome-engineering.

135. Doudna, *Crack in* Creation, xiii.

136. Doudna, *Crack in Creation*, 243–244.

137. Doudna, *Crack in Creation*, 199–200.

138. Yazzie-Lewis and Zion, "Leetso," 2.

139. Yazzie-Lewis and Zion, "Leetso," 7.

140. Yazzie-Lewis and Zion, "Leetso," 8.

141. Yazzie-Lewis and Zion, "Leetso," 10.

142. See *Village of Widows*, dir. Peter Blow (1999, Lindum Films, Peterborough, Ontario, Canada), DVD. At an "Earth Honoring Faith" event at Ghost Ranch in New Mexico, Mindahi C. Bastida Munoz, the Director of the Original Caretakers Program, Center for Earth Ethics at Union Theological Seminary, spoke of a delegation of Indigenous peoples who traveled to the site of the Fukishima meltdown to make offerings and reparation.

143. Doudna, *Crack in Creation*, 246.

144. Howard Ashman (lyricist) and Alan Mencken (composer), "Mean Green Mother from Outer Space," written for the film *Little Shop of Horrors*, dir. Frank Oz (1986; Burbank, CA, Warner Brothers), film.

145. John Collier, "Green Thoughts," in *The John Collier Reader* (New York: Alfred A. Knopf, 1972), 385–399.

146. *Little Shop of Horrors*, directors cut, dir. Frank Oz (2012; Warner Video), DVD.

147. Val Plumwood, "Being Prey," in *The New Earth Reader: The Best of Terra Nova*, ed. David Rothenberg and Maria Ulvaeus (Cambridge, MA: MIT Press, 1999), 76–91, esp. 90.

Chapter 7

1. Geneva Smitherman, *Black Talk*, Revised Edition, 301. See also Geneva Smitherman, *Word from the Mother: Language and African Americans* (New York: Routledge, 2006).

2. Geneva Smitherman, *Black Talk: Words and Phrases from the Hood to the Amen Corner* (Boston: Houghton Mifflin, 1994), 7.

3. Bambara, "What It Is I Think I'm Doing Anyhow," 325.

4. Smitherman, *Black Talk*, Revised Edition, 133.

5. Geneva Smitherman, *Talkin and Testifyin: The Language of Black America* (New York: Houghton Mifflin, 1977), 60. Smitherman stresses that the word "*never* refers to a person who has sex with his mother.".

6. Brittany Spanos, "Janelle Monáe Frees Herself," *Rolling Stone*, April 26, 2018, https://www.rollingstone.com/music/music-features/janelle-monae-frees-herself-629204/.

7. Curtis Mayfield, "Doo Wop Is Strong in Here," from the 1977 film *Short Eyes*, https://www.letras.mus.br/curtis-mayfield/380542/.

8. Fred Moten, "Blackness and Nothingness (Mysticism in the Flesh)," *South Atlantic Quarterly* 112, no. 4 (Fall 2013): 737–780, esp. 738.

9. Marlo D. David, *Mama's Gun: Black Maternal Figures and the Politics of Transgression* (Columbus: Ohio State University Press, 2016), 7.

10. "About Us," *Mutha* magazine, http://muthamagazine.com/about/.

11. Minh Nguyen, "An 'Imaginary Museum' at Art Gallery Enlivens Queer and Trans History," *Seattle Weekly*, May 31, 2017, https://www.seattleweekly.com/arts/a-qtrans-hirstory-in-99-objects/.

12. Minita Ghandi, *Muthaland*, 2017, http://www.minitagandhi.com/theatre/.

13. Cheryl Strayed, *Tiny Beautiful Things: Advice on Love and Life from Dear Sugar* (New York: Vintage Books, 2012), 60.
14. Cathy J. Cohen, "Punks, Bulldaggers, and Welfare Queens: The Radical Potential of Queer Politics?" *GLQ: A Journal of Lesbian and Gay Studies* 3 (1997): 437–465, esp. 437.
15. Gloria E. Anzaldúa, "Now Let Us Shift . . . the Path of Conocimiento . . . Inner Work . . . Public Acts," in *This Bridge We Call Home: Radical Visions for Transformation*, ed. Gloria E. Anzaldúa and AnaLouise Keating (New York: Routledge, 2002), 540–570, esp. 558.
16. Nancy Hartsock, "Foucault on Power: A Theory for Women," in *Feminism/Postmodernism*, ed. Linda J. Nicholson (New York: Routledge, 1990), 157–175, esp. 161.
17. Rachel Rosenzweig, *Worshipping Aphrodite: Art and Cult in Classical Athens* (Ann Arbor: University of Michigan Press, 2004), 14. Natalie King-Pedroso, "'I-Tie-All My People-Together': New World Appropriations of the Yoruba Deity Oshun in Toni Morrison's *Tar Baby* and Derek Walcott's *Omeros*," *Journal of Caribbean Studies* 15 (2000): 61–93.
18. Audre Lorde, "Uses of the Erotic: The Erotic as Power," *Sister Outsider* (Trumansburg, NY: Crossing Press, 1984), 53–59, esp. 56.
19. Paula Gunn Allen (1939–2008) first suggested this to me, probably in about 1995 in Albuquerque, New Mexico.
20. Harris, *Ecowomanism*, 26–27.
21. Awiakta, *Selu*, 9.
22. June Jordan, "The Creative Spirit: Children's Literature," in *Revolutionary Mothering: Love on the Front Lines*, ed. Alexis Pauline Gumbs, China Martens, and Mai'a Williams (Oakland, CA: PM Press, 2016), 11–18, esp. 11-12.
23. See Andrea O'Reilly, *Matricentric Feminism: Theory, Activism, and Practice* (Bradford, CAN: Demeter Press, 2016);
24. Jordan, "Creative Spirit," 11–18, esp. 12.
25. Deborah Bird Rose, "Arts of Living on a Damaged Planet," *Love at the Edge of Extinction* (blog), May 17, 2014, http://deborahbirdrose.com/tag/anthropocene/.
26. Thom Van Dooren and Deborah Bird Rose, "Lively Ethography: Storying Animist Worlds," *Environmental*

Humanities 8, no. 1 (2016): 77–94, https://doi-org.ezproxy.fau. edu/10.1215/22011919-3527731.

27. Haraway, *Staying with the Trouble*, 40.
28. Serenella Iovino and Serpil Opperman, "Introduction: Stories Come to Matter," in *Material Ecocriticism*, ed. Serenella Iovino and Serpil Opperman (Bloomington: Indiana University Press, 2014), 8.
29. Harold Bayley, *The Lost Language of Symbolism*, vol. 1: *The Origins of Symbols, Mythologies and Folklore* (1912; New York: Barnes and Noble, 1996), 190.
30. Jack Zipes, *When Dreams Came True: Classical Fairy Tales and Their Tradition* (New York: Routledge, 1999).
31. Bayley, *Lost Language of Symbolism*, 1:190.
32. Bayley, *Lost Language of Symbolism*, 1:194.
33. Caitlín Matthews, *Sophia Goddess of Wisdom: The Divine Feminine from Black Goddess to World Soul* (London: Mandala, 1991), 347.
34. Bayley, *Lost Language of Symbolism*, 190.
35. Margaret Atwood, "Perfect Storms: Writing *Oryx and Crake*," http://www.oryxandcrake.co.uk/perfectstorm.asp.
36. The Elephant Listening Project, Cornel Lab, Cornell University, Ithaca, NY, http://www.birds.cornell.edu/brp/elephant/about/about.html.
37. Katy Payne, "In the Presence of Elephants and Whales," interview with Krista Tippet, *On Being*, National Public Radio. Aug. 13, 2015, https://onbeing.org/programs/katy-payne-in-the-presence-of-elephants-and-whales/.
38. Dreams are a way of learning" in Anishinaabe cultures," Leanne Simpson, "Stories, Dreams, and Ceremonies: Anishinaabe Ways of Learning," *Tribal College: Journal of American Indian Higher Education* 11, no. 4 (2000): https://tribalcollegejournal.org/stories-dreams-ceremonies-anishinaabe-ways-learning/.
39. Kim TallBear, "An Indigenous Reflection on Working beyond the Human/Not Human," *GLQ: A Journal of Gay and Lesbian Studies* 21, no. 2–3 (2015): 230–235, esp. 234.
40. Smitherman, "Who Yo Daddy," *Black Talk*, Revised Edition, 297.
41. "Who's Your Daddy," Wikipedia, https://en.wikipedia.org/wiki/Who%27s_your_daddy%3F_(phrase).

42. Juana María Rodríguez, *Sexual Futures, Queer Gestures, and Other Latina Longings* (New York: New York University Press, 2014), 181.
43. I consider incest in relation to nuclear threat in Caputi, *Gossips, Gorgons and Crones*, 99–140.
44. Allen, *Sacred Hoop*, 209.
45. Allen, *Sacred Hoop*, 209–210.
46. Editorial Board, "A Woman's Rights," *New York Times*, Dec. 28, 2018.
47. David, *Mama's Gun*, ix.
48. Roberts, *Killing the Black Body*, 56–159 and throughout. Adam Cohen, *Imbeciles: The Supreme Court, American Eugenics, and the Sterilization of Carrie Buck* (New York: Penguin Press, 2016). Ladelle McWhorter, *Racism and Sexual Oppression in Anglo-America: A Genealogy* (Bloomington: Indiana University Press, 209), 280–293.
49. Roberts, *Killing the Black Body*, 56–149 and throughout.
50. DeNeen L. Brown, "'Barbaric': America's Cruel History of Separating Children from Their Parents," *Washington Post*, May 31, 2018,
51. Native Youth Sexual Health Network and Women's Earth Alliance, *Violence on the Land, Violence on Our Bodies*, 18–25.
52. Roni Caryn Rabin, "Huge Racial Disparities Found in Deaths Linked to Pregnancy," *New York Times*, May 7, 2019.
53. Katsi Cook, cited in Native Youth Sexual Health Network and Women's Earth Alliance, *Violence on the Land, Violence on Our Bodies*, 20.
54. Loretta J. Ross and Rickie Solinger, *Reproductive Justice: An Introduction* (Oakland: University of California Press, 2017), 9.
55. Lee Maracle, "Decolonizing Native Women," in *Daughters of Mother Earth: The Wisdom of Native American Women*, ed., Barbara Alice Mann (Westport, CT: Praeger, 2006), 29–51.
56. "Our Story," Little Lobbyists, https://littlelobbyists.org/our-leadership.
57. Daly, *Beyond God the Father*, 43. Daly was still using "god" language in 1973 when she wrote this definition of divinity, but subsequently stopped using it, preferring *Goddess*.

58. Mai'a Williams, "Introduction," *Revolutionary Mothering*, 1–2, esp. 2.

59. Alexis Pauline Gumbs, "M/other Ourselves: A Black Queer Feminist Genealogy for Radical Mothering," *Revolutionary Mothering*, 19–31, esp. 20-21.

60. Soraya Chemaly, *Rage Becomes Her: The Power of Women's Anger* (New York: Simon and Schuster, 2018), 91–119.

61. Jordan, "Creative Spirit," 12.

62. Irene Lara, "Sensing the Serpent in the Mother, Dando a Luz la Madre Serpiente: Chicana Spirituality, Sexuality, and Mamihood," in *Fleshing the Spirit: Spirituality and Activism in Chicana, Latina, and Indigenous Women's Lives*, ed. Elisa Facio and Irene Lara (Tucson: University of Arizona Press, 2014), 113–134, esp.115.

63. Kim TallBear, "Making Love and Relations beyond Settler Sexualities," video of a talk sponsored by Ecologies of Social Justice and the Social Justice Institute, University of British Columbia, Vancouver, BC, Canada, Feb. 24, 2016, https://www.youtube.com/watch?v=zfdo2ujRUv8.

64. Patrick Johnson, *Appropriating Blackness: Performance and the Politics of Authenticity* (Durham, NC: Duke University Press, 2003), 91.

65. Johnson, *Appropriating Blackness*, 93–94.

66. Naranjo and Swentzell, "Healing Spaces," 260.

67. Allen, *Sacred Hoop*, 209.

68. Tuhiwai Smith, *Decolonizing Methodologies*, 77.

69. Jeanette Armstrong, "Sharing One Skin," *Cultural Survival Quarterly Magazine*, Dec. 2006, https://www.culturalsurvival.org/publications/cultural-survival-quarterly/sharing-one-skin.

70. Caputi, *Goddesses and Monsters*, 18.

71. Garrard, *Brunelleschi's Egg*, 286–289.

72. Nicole Pasulka and Brian Ferree, "Unearthing the Sea Witch," *Hazlitt*, Jan. 24, 2016, https://hazlitt.net/longreads/unearthing-sea-witch.

73. C. C. Copper, *An Illustrated Encyclopaedia of Traditional Symbols* (London: Thames & Hudson, 1978), 176.

74. David Quammen. "The Scientist Who Scrambled Darwin's Tree of Life," *New York Times*, Aug. 13, 2018.

75. Kate Detwiler, conversation with the author, August, 24, 2018. See also J. A. Hart, K. M. Detwiler, C. C. Gilbert, A. S. Burrell, J. L. Fuller et al., "Lesula: A New Species of Cercopithecus Monkey Endemic to the Democratic Republic of Congo and Implications for Conservation of Congo's Central Basin," *PLoS ONE* 7, no. 9 (2012): e44271, doi:10.1371/journal.pone.0044271

76. darkdaughta, "Mother Earth Tree of Life," *Creative Resistance*, n.d., http://www.creativeresistance.org/mother-earth-tree-of-life/.

77. David Suzuki and Wayne Grady, *Tree: A Life Story* (Berkeley, CA: Greystone Books, 2004), 68.

78. Christel N. Temple, "The Emergence of Sankofa Practice in the United States: A Modern History," *Journal of Black Studies* 41, no. 1 (2010): 127–150.

79. Temple, "Emergence," 130.

80. Gumbs, "M/other Ourselves," 21–22.

81. "Interview with Farmer and Healer Amber Tamm Canty," Inside the Rift, May 8, 2017, https://www.insidetherift.net/metaphysics/2017/5/8/vhohxt7lpwnsmkk8t9q0awwji5x8u8.

82. See, for example, Raine, *After Silence*; Charlotte Pierce-Baker, *Surviving the Silence: Black Women's Stories of Rape* (New York: W. W. Norton, 1998); and Eve Ensler, *Vagina Monologues* (New York: Villard, 1998).

83. Lee Hester and Jim Cheney, "Truth and Native American Epistemology," *Social Epistemology* 15, no. 4 (2001): 319–334, esp. 326.

84. Barbara Christian, "The Highs and Lows of Black Feminist Criticism (1990)," in *Feminisms: An Anthology of Literary Theory and Criticism*, ed. Robyn R. Warhol and Diane Price Herndl (New Brunswick, NJ: Rutgers University Press, 1997), 51–56, esp. 56.

85. Hildegard of Bingen, *Book of Divine Works with Letters and Songs*, ed. Matthew Fox (Santa Fe, NM: Bear & Co., 1986), 114.

86. Marta Weigle, *Creation and Procreation: Feminist Reflections on Mythologies of Cosmogony and Parturition* (Philadelphia: University of Pennsylvania Press, 1989).

87. The *Oxford English Dictionary* added *cunty* in 2014, defining it as: "Of, relating to, or from the vagina; resembling, suggestive

of, or associated with the female genitals; Despicable; highly unpleasant; extremely annoying."

88. Christian, "Highs and Lows," 56.

89. Jessica Benjamin, *The Bonds of Love: Psychoanalysis, Feminism, and the Problem of Domination* (New York: Pantheon, 1988), 124.

90. Jane Caputi, "'Cuntspeak': Words from the Heart of Darkness," in *Not For Sale: Feminists Resisting Prostitution and Pornography*, ed. Christine Stark and Rebecca Whisnant (Melbourne, AUS: Spinifex Press, 2004), 362–385.

91. Andrea Dworkin, *Intercourse* (New York: Free Press, 1987).

92. Maureen Mullarkey, "Hard Cop, Soft Cop," *The Nation*, May 30, 1987, 721–726.

93. Ginger Adams Otis, "*Chienne de Farge*: French Feminists Vigilant against Sexist Language," *Sojourner: The Women's Forum* (Jan. 2000): 9.

94. Jane Caputi, "From (Castrating) *Bitch* to (Big) *Nuts*: Genital Politics in 2016 Election Campaign Paraphernalia," in *Bad Hombres and Nasty Women*, ed. Christine Kray, Tamar Carroll, Hinda Mandell (Rochester, NY: University of Rochester Press, 2018), 26–41, esp. 28.

95. Ben Dowell, "Mary Beard Suffers 'Truly Vile' Online Abuse after Question Time," *The Guardian*, Jan. 21, 2013, https://www.theguardian.com/media/2013/jan/21/mary-beard-suffers-twitter-abuse.

96. Brittney Cooper, "Pussy Don't Fail Me Now," *Crunk Feminist Collective*, Jan. 23, 2017, http://www.crunkfeministcollective.com/2017/01/23/pussy-dont-fail-me-now-the-place-of-vaginas-in-black-feminist-theory-organizing/.

97. This *Hustler* self-promotion (from around 1976) is described in Diana E. H. Russell, *Dangerous Relationships: Pornography, Misogyny, and Rape* (Thousand Oaks, CA: Sage Publications, 1998), 65.

98. Mircea Eliade, *The Forge and the Crucible: The Origins and Structures of Alchemy*, 2nd ed., trans. Stephen Corrin (Chicago: University of Chicago Press, 1978), 21.

99. Sarah Iles Johnston, *Religions of the Ancient World: A Guide* (Cambridge, MA: Harvard University Press, 2004), 384.

100. Charles Segal, "The Gorgon and the Nightingale: The Voice of Female Lament and Pindar's Twelfth Pythian Ode," in *Embodied Voices: Representing Female Vocality in Western Culture*, ed. Leslie C. Dunn and Nancy A. Jones (New York: Cambridge University Press, 1994), 17–34, esp. 18.

101. Sigmund Freud, "Medusa's Head (1940 [1922])," in *Standard Edition of Complete Psychological Works of Sigmund Freud*, vol. 18 (London: Hogarth Press, 1953), 273–274.

102. Segal, "Gorgon and the Nightingale," 30.

103. Glenys Livingstone, *PaGaian Cosmology: Re-inventing Earth-based Goddess* (New York: iUniverse, 2005), 66.

104. Caputi, "Naked Goddess," in *Goddesses and Monsters*, 370–386.

105. Laura S. Grillo, *An Intimate Rebuke: Female Genital Power in Ritual and Politics in West Africa* (Durham, NC: Duke University Press, 2018), 2.

106. Carolee Schneemann, *Imaging Her Erotics: Essays, Interviews, Projects* (Cambridge, MA: MIT Press, 2002), 299.

107. Pérez, "Irigaray's Female Symbolic," 114.

108. Clarissa Pinkola Estés, *Women Who Run with the Wolves: Myths andStories of the Wild Woman Archetype* (New York: Ballantine Books, 1992), 340.

109. Annie Sprinkle, *Post-Porn Modernist: My 25 Years as a Multimedia Whore* (San Francisco: Cleis Press, 1998), 165.

110. Anne Fausto-Sterling, *Sex/gender: Biology in a Social World* (New York: Routledge, 2012).

111. Allen, *Sacred Hoop*, 268.

112. *Matrix* is a Latin word meaning "uterus" and means the same in English, as well as "a place or medium in which something is originated, produced, or developed; the environment in which a particular activity or process begins; a point of origin and growth" (*OED*).

113. Mary Daly, *Gyn/Ecology: The Metaethics of Radical Feminism* (Boston: Beacon Press, 1978). Daly first here and then throughout her work offers a philosophy of "the Background" as source of connection, the context of creation throughout the Universe.

114. David, *Mama's Gun*, 108.

115. Roberts, *Killing the Black Body*, 45.

116. Cooper, "Pussy Don't Fail Me Now."
117. Philip, "Dis Place," 100-101.
118. Philip, "Dis Place," 77, 77, 101, 78, 107, 78.
119. Philip, "Dis Place," 107
120. Lightning is a complex symbol of revelation, creation, and destruction, also often signifying male fertilizing energy. Chevalier and Gheerbrant, *Penguin Dictionary of Symbols*, 607.
121. Favianna Rodriguez, "Don't Frack with Me," 2017. Rodriguez makes this statement on her Facebook page, https://www.facebook.com/faviannaart/photos/a.10150569444991110/10156554961631110/?type=1&theater.
122. Cynthia Eller criticizes feminist reclamation of many goddess images as perpetuating "a long-standing tradition of objectifying women and reducing them to their sexual parts." Cynthia Eller, "Divine Objectification: The Representation of Goddesses and Women in Feminist Spirituality," *Journal of Feminist Studies in Religion* 16, no. 1 (2000): 23–44, esp. 30.
123. Vine Deloria, Jr., "If You Think About It, You Will See That It Is True," in *Spirit and Reason: The Vine Deloria, Jr., Reader*, ed. Barbara Deloria, Kristin Foehner, and Sam Scinta (Golden, CO: Fulcrum, 1999), 40–60, esp. 50–51. This same law of respect underlies the story of Selu, the Cherokee Corn-Mother. Awiakta, *Selu*, 21–22, 252, and throughout.
124. Cooper, *Eloquent Rage*, 23.
125. Cooper, *Eloquent Rage*, 26.
126. Cooper, "Pussy Don't Fail Me Now."
127. Philip, "Dis Place," 78.
128. Caputi, "Cunctipotence."
129. Stephen Greenblatt, "How St. Augustine Invented Sex," *New Yorker*, June 19, 2017.
130. Cited in Rosemary Agonito, *History of Ideas on Woman: A Source Book* (New York: Berkeley, 1977), 4.
131. Paul Joannides, *Guide to Getting It On!* (Waldport, OR: Goofy Foot Press, 2000), 13.
132. James Nelson, "Embracing Masculinity," in *Sexuality and the Sacred: Sources for Theological Reflection*, ed. James B. Nelson and Sandra P. Longfellow (Louisville, KY: Westminster John Knox Press, 1994), 195–215, esp. 206.
133. farajajé-jones, "Holy Fuck," 335.

134. Val Plumwood, *Feminism and the Mastery of Nature* (New York: Routledge, 1993), 41.
135. Barbara G. Walker, "Cunt," in *The Woman's Encyclopedia of Myths and Secrets* (Edison, NJ: Castle Books, 1983), 197–198, esp. 197.
136. Unfortunately, I do not have a source for this image, just a slide.
137. Frédérique Apffel-Marglin, "Yoni," in *Encyclopedia of Religion*, ed. Lindsay Jones, 2nd ed. (Detroit, MI: Macmillan Reference USA, 2004), 9905–9909, esp. 9905.
138. Amber Tamm Canty, http://ambertamm.com/.
139. Walker, *Color Purple*, 190–191.
140. "Not fatherless ... How about motherful?" This statement from Khadijah Martin is quoted in Gumbs, "M/other Ourselves," 29.
141. Virginia Woolf, "A Sketch of the Past," in *Moments of Being*, 2nd ed., ed. Jeanne Schulkind (San Diego: Harcourt Brace Jovanovich, 1985), 64–159, esp. 71–72. .

Chapter 8

1. Hogan, *Dwellings*, 85.
2. Val Plumwood, "Tasteless: Towards a Food-Based Approach to Death," *Environmental Values* 17, no. 3 (2008): 323–330.
3. Darryl Pinckney, "Opinions and Poems," *New York Times*, Aug. 9, 1981.
4. Jordan, "Getting Down to Get Over," 175.
5. Jordan, "Getting Down to Get Over," 173–174.
6. Jordan, "Getting Down to Get Over," 180–183.
7. Jordan, "Getting Down to Get Over," 182.
8. Rickey Vincent, *Funk: The Music, the People, and the Rhythm of the One* (New York: St. Martin's Griffin, 1996), 4.
9. Carson, "To Understand Biology," in Lear, *Lost Woods*, 192–194, esp. 193.
10. John Thomas Biggers, *Web of Life*, Casein with egg emulsion on canvas, 72 x 312, 1960. Texas Southern University Art Gallery.
11. John Thomas Biggers, in *Black Arts in Huston*, cited in Olive Jensen Theisen, *Walls That Speak: The Murals of John Thomas Biggers*, updated ed. (Denton: University of North Texas Press, 2010), 55.

12. Wardlaw, "Metamorphosis," in *Art of John Biggers*, 46.

13. Wardlaw, "Metamorphosis," 47.

14. Wardlaw, "Metamorphosis," 47.

15. John Biggers, *Ananse: The Web of Life in Africa* (Austin: University of Texas Press, 1982), 28.

16. Biggers, *Ananse*, 28.

17. Biggers, *Ananse*, 27.

18. Biggers, *Ananse*, 29.

19. Alison de Lima Greene, "John Biggers: American Muralist," in Wardlaw, *Art of John Biggers*, 96–107, esp. 103.

20. Joanne M. Braxton, "Ancestral Presence: The Outraged Mother Figure in Contemporary Afra-American Writing," in *Wild Women in the Whirlwind: Afra-American Culture and the Contemporary Literary Renaissance*, ed. Joanne M. Braxton and Andrée Nicola McLaughlin (New Brunswick, NJ: Rutgers University Press, 1990), 299–315, esp. 301.

21. Theisen, *Walls That Speak*, 57.

22. Biggers, *Ananse*, 28.

23. Ytasha L. Womack, *Afrofuturism: The World of Black Sci-Fi and Fantasy Culture* (Chicago: Lawrence Hill Books, 2013), 9.

24. Womack, *Afrofuturism*, 103.

25. Ken Johnson, "'Anthems for the Mother Earth Goddess' at Andrew Edlin Gallery," *New York Times*, July 23, 2015.

26. Kevin Sampson, email correspondence, March 30, 2016.

27. These refer to killings of Black men. The #SAYHERNAME movement "calls attention to police violence against Black women, girls and femmes, and demands that their stories be integrated into calls for justice, policy responses to police violence, and media representations of police brutality." See http://aapf.org/shn-campaign.

28. Pablo Picasso, *Woman in Green (Dora)*, oil on canvas, 130.0 x 97.0 cm. Fondation Beyeler, Riehen/Basel, Beyeler Collection, 1944, https://www.fondationbeyeler.ch/sammlung/werk/detail/191-femme-en-vert-dora/.

29. Chevalier and Gheerbrant, *Penguin Dictionary of Symbols*, 15–17.

30. Kevin Sampson, email correspondence, March 30, 2016.

31. Butler, *Parable of the Talents*, 24. Before Trump deployed it, the phrase had been used by Ronald Reagan.

32. Octavia E. Butler, *Parable of the Sower* (New York: Four Walls Eight Windows, 1993), 13–14.

33. Butler, *Parable of the Talents*, 313.

34. Butler, *Parable of the Sower*, 3.

35. Butler, *Parable of the Sower*, 17.

36. Butler, *Parable of the Talents*, 50.

37. Gleason, *Oya*, 11

38. Gleason, *Oya*, 3–4.

39. Randy P. Conner, with David Hatfield Sparks, *Queering Creole Spiritual Traditions: Lesbian, Gay, Bisexual, and Transgender Participation in African-Inspired Traditions in the Americas* (New York: Harrington Park Press, 2004), 74–75.

40. Gleason, *Oya*, 2.

41. Alfred Huang, *The Complete I Ching* (Rochester, VT: Inner Traditions, 1998), 26.

42. Wangechi Mutu, *End of Eating Everything*, 2013. Animated color video with sound, 8 minutes, 10-second loop. Copyright Wangechi Mutu. Gladstone Gallery, New York and Brussels, and Victoria Miro Gallery, London. Commissioned by the Nasher Museum of Art at Duke University, Durham, NC.

43. "Wangechi Mutu + Santigold interview," Nasher Museum of Art at Duke University, MOCAtv, March 21, 2013, https://www.youtube.com/watch?v=XczdrcLUxMA.

44. Mary Daly, *Quintessence; Black Panther*, dir. Ryan Coogler (2018; Burbank, CA: Marvel Studios), film.

45. "Black Earth Ensemble," website, https://www.nicolemitchell.com/black-earth-ensemble.

46. Nicole M. Mitchell, "Staircase Struggle" (liner notes) on *Mandorla Awakening II: Emerging Worlds*, 2017, FPE Records, track 7, compact disc.

47. Nicole Mitchell, "Interview," *Wire*, April 2017, https://www.thewire.co.uk/audio/tracks/listen-to-nicole-mitchell-s-mandorla-awakening.

48. Mitchell, "Staircase Struggle."

49. Mitchell, "Interview."

50. Mitchell, "Interview."

51. Mitchell, "Staircase Struggle."

52. Nicole Mitchell, quoted on FPE Records website, http://www. fperecs.com/product/mandorla-awakening-ii-emerging-worlds/.

53. Chevalier and Gheerbrant, *Penguin Dictionary of Symbols*, 16.

54. Chevalier and Gheerbrant, *Penguin Dictionary of Symbols*, 16.

55. Riane Eisler, *The Chalice and the Blade: Our History, Our Future* (San Francisco: Harper & Row, 1987).

56. Maggie Macnab, *Decoding Design: Understanding and Using Symbols in Visual Communication* (Cincinnati, Ohio: HOW Books, 2008), 59–61.

57. Mitchell, "Staircase Struggle."

58. *Frozen*, dir. Chris Buck (2013; Burbank, CA: Walt Disney Animation Studios), film.

59. *Moana*, dir. Ron Clements and John Musker (2016; Burbank, CA: Walt Disney Animation Studios), film.

60. *Shrek*, dir. Andrew Adamson and Vicky Jenson (2001; Glendale, CA: DreamWorks Animation), film.

61. Hogan, *Power*; Jesmyn Ward, *Salvage the Bones* (New York: Bloomsbury, 2011),

62. Ekkehart Malotki and Michael Lomatuway'ma, *The Magic Hummingbird: A Hopi Folktale*, illustrated by Michael Lacapa (Walnut, CA: Kiva Publishing, 1996).

63. Amaterasu retreats after her brother Susan-o "pierced her vagina with a spindle-shaft." Shahrukh Husain, *The Goddess* (Boston: Little, Brown, 1997), 65.

64. Winifred Milius Lubell, *The Metamorphosis of Baubo: Myths of Women's Sexual Energy* (Nashville, TN: Vanderbilt University Press, 1994), 14.

65. Grillo, *Intimate Rebuke*.

66. Zora Neale Hurston, *Tell My Horse* (Berkeley: Turtle Island, 1983, 1938), 137.

Coda

1. This 2007 dream encounter took place in Bandelier National Monument in New Mexico, where I was standing in one of the carved-out cliff dwellings of the ancient Indigenous peoples of the

region. This being named himself as "William."

2. Robin Wall Kimmerer, "Returning the Gift," Center for Humans and Nature, Oct. 1, 2013. Kimmerer has a similar inclination, writing, "Earth is asking for our vote." https://www.humansand-nature.org/earth-ethic-robin-kimmerer. This is an earlier version of Kimmerer's 2014 essay "Returning the Gift."

INDEX

12 Years a Slave (film and novel), 81
2001: A Space Odyssey (film), 115
2SQ (Two Spirit and queer), 11, 84,
    248n26. *See also* two spirit

ableism, 29, 31, 196
abuelita knowledge, 110, 138
Ackerman, Diane, 43–44, 62
Adam and Eve, 46, 62, 73, 137, 154,
    191. *See also* Eve; Genesis
Adams, Henry, 125–126
African slave trade, 107. *See also*
    slavery
Africana religion and spirituality, 25,
    30, 85, 159, 191, 219, 220–221,
    228, 236
Afrofuturism, 20, 223
Albert, Gerrard, 68
alchemy, 161, 165
Alexander VI, Pope, "Inter
    Caetera," 52
Alim, Samy H., 33

Allen, Paula Gunn, 60, 72, 84, 113,
    126, 184, 189, 194
Amaterasu, 235
"An American Dream of Venus,"
    133–136
Ananse (Anansa), 221, 233
André 3000, 34
Andrews, Tamra, 7
animality, 29, 256n20
animals, 24, 71, 73, 95, 125, 137,
    158, 165, 176, 183
    as beings, 5, 11, 16, 170, 188
    as creative and inventive, 56
    dominated and tortured by humans,
        47, 49, 56, 78, 83–84, 92, 95, 111,
        135, 137, 158, 165, 191, 225
    humans as, 29
    wild, 1, 210
Anishinaabeg philosophy, 14, 147
Annihilation (novel and film), 176
Anthropocene
    as Age of *Man*, 77

Anthropocene (*cont.*)
  as Age of the
    Motherfucker, 5, 108
  as matricidal and deicidal
    assault, 31
  defined and described, 3, 9, 11–
    13, 15–16, 19, 24, 42–51, 55–62
  discourse of, 4, 15–16, 19, 24, 42–
    44, 48, 50–51, 55–62
  rapism and, 4
  script of, 183
  spiritual meanings of, 180
  undoing of, 197
anti-blackness, 114, 120, 130
Anzaldúa, Gloria, 30, 183
Aphrodite. See *Venus*
Apple computer logo, 136–137
Aristotle, 211
Armstrong, Jeanette, 194–195
Arrival, The (film), 169
artificial intelligence (AI), 62,
  117, 128
Artistes Indigenos Graphics, 213
Atabey da la Luz al Coquí
  (Torres), 161
Atwood, Margaret, 82, 187
Aunt Jemima, 122–123
Awiakta, Marilou, 2, 37, 39, 68, 144,
  312n123

Bacon, Francis, 29, 46
Badu, Erykah, 204
Baginski, Max, 17, 216
Bambara, Toni Cade, 11, 100,
  181, 240
Baptist, Edward E., 108–109
Barbarella (film), 115
Barthes, Roland, 140
Bartholomew and the Oobleck
  (Seuss), 175–176
basic white, 113–124

Baubo, 236
Beauvoir, Simone de, 125
Beckert, Sven, 109
Bee, Tom, 110
Beloved (Morrison), 99, 115, 159
Benjamin, Jessica, 199
Berry, Daina Ramey, 87
Bierman, Judith, 102–103
Big Baby (Fitz), 58–60, 63, 212
Biggers, John Thomas, 220–223
biohackers, 64
biology, defined by Rachel
  Carson, 219
biophobia, 31
Birth of Venus (Botticelli), 134
Black English, 5, 10, 18, 25, 27, 34,
  146, 147, 169, 181, 213. See also
  Black Talk
Black feminism, 25, 32, 106, 199,
  200, 211
#BlackLivesMatter, 35, 225
Black Madonna, 121–126, 233
Black Panther (film), 230
black space, 115
Black Talk (Smitherman), 3, 5,
  181–182
blackness, 120
  coded as evil, 114
  blue, 111–112, 117, 127, 137,
    145, 196
Bluest Eye (Morrison), 114
Bordo, Susan, 57
Brand, Steward, 50
Braxton, Joanne, 221
Brennan, Teresa, 58, 102, 118
Brontë, Emily, 21
Brown, James, 34
Brown, Michael, 224
Buber, Martin, 158
Burke, Tarana, 103
Butler, Octavia, 227–229, 232

Cáceres, Berta, 69–74
Cajete, Greg, 17, 21, 23, 119, 164
calling the "Mutha," 4, 37–38
Canty, Amber Tamm, 197, 213–215
capitalism, 46, 60, 84, 87, 104,
    107–109, 192
  consumer, 4, 25
  corporate, 60
  global, 104, 108–109, 234
  motherfucking as paradigm for,
    108–109
  rapacious, 70, 87, 102
Capitalocene, 87, 88, 102, 109
Carson, Rachel, 15, 42, 45, 48, 73,
    76, 117, 192, 219
ceremony, 39, 66, 75, 133, 151, 153,
    176, 216, 229, 236, 240
Chapman, Tracy, 28
Cheney, Jim, 98
Chiang, Ted, 169
Chicana feminists, 203
Chippewa history, 23
Christian, Barbara, 198–199
chromophobia, 115
Church, George, 60
Cinderella, 186–187
climate change, 3, 9, 43, 64, 66–68,
    140, 170, 171, 173
  as evidence of changing Earth
    patterns, 173
  as the response of the "Mutha,"
    184, 227, 229
  eco-fascism and, 15
  racism and, xi, 19, 35
Clinton, Hillary, misogynist slurs
    against, 200
clitoris, 152, 204, 212
Cochrane, Rachel, 18
cognitive justice, 20
Cohen, Cathy, 183
colonialism, 4, 69, 118–121, 138

as rape, 78, 85–85
distortion of Indigenous
    knowledge and systems,
    84–85, 95
  settler, 19, 20, 75–76, 78
color (meanings of), 110–115
  synthetic, 130–137. *See also*
    individual colors
Color Purple, The (Walker), 214
colorlessness, 110, 113, 130, 133,
    141, 153
Columbus, Christopher, 28, 61,
    78, 122
consciousness
  green, 156, 169
  human, 65, 135
  other than human, 24, 128, 170
consumerism, 4, 13, 25, 42, 118,
    126, 230
convocation, 240
Cook, Tekatsi' tsiah:khwa Katsi, 94
Cooper, Brittney, 200, 207, 210
COPINH–The Civic Council of
    Popular AND indigenous
    Organizations of Honduras,
    70, 73, 74
Corn Blossom (Margaret
    Tafoya), 167
Cowan, Liza, 36, 148
creative spirit, 185, 192
creativity, 32, 72, 139, 165, 185, 223
CRISPR, 177, 178, 179
Crutzen, Paul, 13, 46, 48
Cullors, Patrisse, 35, 36
cunctipotence, 142, 240
  defined and illustrated, 211,
    213–215
cunt, 31, 32, 105–106, 132, 136, 198,
    206, 208, 211
  as mouth, 199–200
  defined, 27

cunt (*cont.*)
Earth as, 31, 136, 208
historical context, 28–29
used as a slur, 27, 210, 211
word origins, 212
cunt envy, 135. *See also* womb envy
cuntface, 201
cuntspeak, 200–201, 203, 208,
210, 228
curse, 34, 40–41, 142, 203
cyanobacteria, 139, 145

Daly, Mary, 16, 28, 52, 61, 191, 230
Dambalah and Ayida Wèdo, 156
Danaylov, Nikola, 64
Danbala, 112
dark and darkness, 29–30
darkdaughta, 196
Darwin, Charles, 195
David, Marlo D., 182, 204
Davies, Jeremy, 12, 62
da Vinci, Leonardo, 114
Davis, Adrienne, 86, 88, 106
Davis, Angela, 86
death, 3, 8, 17, 34, 50, 57, 121, 126,
145, 150, 228
as gift of the Earth, 113, 141
as intrinsic to creation, 3, 17, 75,
88, 112, 120, 126, 145, 159,
165–166, 191
Man's desire to defeat, 62
premature, 3, 11, 113, 240
decolonization, 10, 18, 76
Deer, Sarah, 83, 84, 103
deicide, 16, 40, 48, 49, 76, 195
Deloria, Vine, 22, 210
Delphi, 201
Demeter, 39, 148, 235
Descartes, Rene, 46
Detwiler, Kate, 196
devotion, 46, 49, 57, 66, 150,
184, 239
Dhillon, Jaskiran, 69

Diana of Ephesus, 125–126
Dickens, Karla, 123
dirt, 27, 58, 65, 109, 117, 121, 234
defined, 29
peopled treated like, 31, 117
worship of, 159
dirty, 120–121, 126
as slur, 31, 119–120
opposed to clean, 29–30, 32, 116
words, 31, 236
dirty mind, 26, 32, 198, 208
discredited knowledge, 20, 171
Don't Frack With Me (Rodriguez),
208–209
Don't Frack Your Mother
(Sayles), 101
Doudna, Jennifer, 177, 178, 179
Douglas, Frederick, 122
Douglas, Katie, 98
Dr. Doom, 158
*Dr. Strangelove: How I Learned to
Stop Worrying and Love the
Bomb* (film), 133, 98
Dracula, 152, 237
Dracula (theater poster), 153
dreams and dreaming, 4, 21, 35,
144–145, 147, 178, 187–188, 239
drill, sexualized, 32, 56,
91–94, 100
dualism, hierarchical, 30, 34, 36, 72,
99, 113, 204, 223
DuBois, W. E. B., 115
Duval-Carrié, Edouard, 162, 163
Dworkin, Andrea, 32, 80, 108,
199, 200
Dyke, Marty O., 103

Earth Mother. *See* Mother
Nature-Earth
earthbound, 25–26, 65, 170, 189,
192, 194, 195, 203, 206, 213,
214, 216, 219, 234
earth sounds, 75, 201, 237

eco-ability, 20, 64, 71, 138
eco-apocalypse, 4, 235
   biblical, 51–52
eco-fascism, 15
ecocide, 4, 11, 30, 48, 63, 158
ecofeminism, 14, 66, 77, 94,
   159, 223
Economist (magazine), 55, 139
ecosexuality, 36, 204
ecowomanism, 10, 14, 85,
   185, 223
Eisler, Riane, 233
elephants, 12, 187
English language, 5, 18, 25, 27, 146,
   169, 213
environmental justice, 35, 45, 66,
   70, 97, 98, 190, 191, 192, 223
environmental violence, 83,
   97–98, 232
envy, 8, 13, 135
Erdrich, Louise, 31, 176
erotic, the, 32, 184
erotophobia, 31
essentialism, 18, 71, 210
   and Indigenous
   philosophy, 72–73
Estés, Clarissa Pinkola, 203, 204
Estes, Nick, 69
ethnic cleansing, 30
etiquette, ecological, 176, 180
eugenics, 117, 118, 138, 177
Eurocentrism, 8–9, 18–20, 30,
   83, 232
Eve, 46, 73, 154
"everythang is everything," 147
Expoção International do
   Centenário
extractivism, 118, 120

Fantastic Four, The (film), 157
farajajé-jones, elias, 31, 212
Farrington, Lisa E., 156
Faustine, Nona, 104, 105, 107, 108

feeding the green, 144–145, 147,
   171, 174
fembot, 127, 128, 129, 141
femicide, 108
   of Indigenous women, 96–97
feminism, 71, 97, 207, 210
femme, 5, 29, 36, 37, 133, 135, 191,
   193, 199, 224
fertility goddess, 113, 126
Fitz, W. Grancel, 58, 59
flipping the script, 182, 203
Florida, 88, 244
folklore and fairy tales, 4, 22, 34,
   95, 126, 127, 147, 186, 221
food, 1, 43, 48, 64, 100, 112, 115,
   120, 121, 124, 145, 190, 192
   global agrifood, 140
Forbes, Jack, 23, 24, 78, 174
Fortune (magazine), 131, 133,
   134, 135
Foster, Frances, 87
fountain of life, 144, 148, 150, 156,
   158, 221, 233
fracking, 83, 99, 100, 102, 108, 225
Francis, Pope, 8
Frankenstein, 175, 178
Freud, Sigmund, 13, 135, 201
Frichner, Gonnella, 66, 68
*From Her Body Sprang Their
   Greatest Wealth, Wall Street*
   (Faustine), 104–105
Frozen (film), 234
Fruit of the Poisonous Tree
   (Sampson), 223–227
fuck
   definition, 5, 183
   fusion of sex and violence, 5, 32–34,
   89, 92, 100, 109, 188–189
funk, 20, 30, 32, 34–35, 114, 121,
   133, 135, 146, 198, 234, 236
   in African cosmologic, 219

Gandhi, Minita, 182

Garrard, Mary, 52, 62
Gauguin, Paul, 150
gender, 3, 35, 71, 72, 80, 96, 98, 177,
    189, 194
  alternatives to heteropatriarchal,
    81, 84, 112–113, 177, 190, 211,
    223, 229
  heteropatriarchal gender system,
    72, 80, 81, 84, 97–98, 189
  non-binary, 36, 81, 113, 133, 203
  pronouns, 18, 36
gender-based violence, 14, 29, 80,
    88, 97, 108
gendercide, 29
Genesis (biblical book), 46, 60–61,
    73, 137, 152–153
genital power, 203, 208, 236
genitals, 14, 96, 112, 134, 198–199,
    201, 202, 207, 208, 235–236
  earth, 198, 201, 208. *See also*
    cunt, penis
genocide,
  of Indigenous peoples, 78, 81, 83,
    85, 87, 96, 109, 169
  and gynocide, 169
geoengineering, 137–138
Get Out (film), 72–73
"Getting Down to Get Over"
    (Jordan), 33, 217–219
GFP Bunny (Kac), 166–168
gifts, 65, 73, 151, 172, 176
  of Nature-Earth, 16–17,
    19, 65, 73, 113, 124, 156,
    176, 240
  Pomo concept, 151
  reciprocating, 169, 176, 179, 240
Gilgamesh, 63–64, 82
Gilligan, James, 96–97
Giving Tree, The (book), 151–152
Glave, Dianne D., 151–152
Globopolis, 117, 140, 216
God, 72, 176

  in Butler, 227–228
  in Daly, 191
*God as Architect/Builder/Geometer/
    Craftsman*, 52–53
God, patriarchal, 8, 15–16, 27, 31,
    46, 48–55, 57, 60–62, 80, 115,
    153, 160, 212, 214, 215, 227
  as mask of *Man*, 40, 49
gods
  of the Anthropocene, 42, 45, 48
  humans as, 9, 16, 49–55
God Species, 9, 16, 50–51, 60, 140
Goddexx, 37, 204
Goldman, Emma, 17, 216
Gorgon, 195, 201–202. *See
    also* Medusa
grammar of animacy, 169–171
Great Acceleration, 42, 118,
    126, 220
Greely, Henry T., 129
green, 110–112, 117, 119, 137, 142
  in environmental movement,
    19, 100
  green myth, folk and pop culture
    figures, 147–156
  green and red, 17, 148–152, 156,
    159, 161, 239
Green Belt Movement, 146
green Christ, 150, 233
green consciousness, 156, 169
green fire, 152, 161, 164–172
Green, Nancy, 122–123
greenness. See *viriditas*
Griffin, Susan, 30, 77
Grillo, Laura S., 203
Grinspoon, David, 43–44, 51
Gumbs, Alexis Pauline, 192
gynocide, 28

Haaland, Deb, 85
Hampton, Jerome Gregory, 128
Handmaid's Tale, The (Atwood), 82

Harari, Yuval Noah, 49, 51
Haraway, Donna, 30, 177, 186, 195
Harris, Eric, 50
Harris, Melanie, 185
Harrison, Robert Pogue, 63–64
Hartman, Saidiya, 133–134
heart, 134, 136, 141, 144, 186,
    197, 200
heart knowledge, 111, 115, 170, 240
Hecht, Gabrielle, 48, 140
Hester, Lee, 198
heteropatriarchy, 36, 70, 73, 78,
    80, 82, 84, 102, 140, 197,
    210, 246n4
Highway, Tomson, 97, 112, 136, 211
Hildegard of Bingen, 161, 165, 198
Hobson, Janell, 87, 134
Hogan, Linda, 16, 216, 234
homophobia, 31, 82, 192, 193
Honduras, 31, 82, 192, 193
hooks, bell, 25, 220
Hopi, 235
Horney, Karen, 13
House of Downtown, 34
Huerta, Delores, 88
Hughes, Langston, 123
human, concept of, 3, 10, 11, 13, 25,
    29–30, 58, 120
    Okangan, 194–195
    Pueblo, 146
human exceptionalism, 10, 24,
    44, 49, 55, 63, 139, 171, 179,
    180, 184
Hurston, Zora Neale, 22, 61–63, 72,
    112, 204, 236
Hustler (magazine), 89–91, 120, 200–201

identity, 19, 96, 103, 111, 183, 194–195,
    210, 213
immortality, 49, 62–65, 113,
    130, 158
impudence, 198

Indigenous knowledge systems and
    ontologies, 17–19, 20, 22–23,
    188, 236
individualism, 58, 83, 195, 216
infinity, 22–23, 112, 125, 204, 234
insidiousness, 139–140
integrity, 35, 37, 86, 102, 108, 111,
    112, 160, 164, 167, 189
intelligence
    of Nature-Earth, 19, 91, 126, 137
    of sex organs, 208
International Movement for Mother
    Earth Rights, 9
invocation, 1, 4, 6, 27, 39, 66, 71
Iovino, Serenella, 186, 198
Ishiguro, Hiroshi, 121

Jack the Ripper, 49, 122, 132
Jacobs, Harriet, 104
jamette, 207, 208, 211
Jarman, Derek, 153
Jefferson, Thomas, 107
Johnson, E. Patrick, 193
Jordan, June, 33, 65, 185–186,
    192–193, 217–219, 227

Kac, Eduardo, 166–168
Kahlo, Frida, 148–149
Keller, Evelyn Fox, 170
Keller, Hellen, 111
Kelly, Kevin, 55
Kelly, Suzanne, 121, 126
Kemmerer, Lisa, 10
Khadar, Amir, 37–38
Kimmerer, Robin Wall, 17, 21, 73,
    95, 160, 165, 169–171, 172–173
King, Marin Luther, Jr., 147
Kiplinger's (magazine), 127–129
Klein, Naomi, 15, 147
Kolbert, Elizabeth, 9
Konsmo, Erin Marie, 75
Kripal, Jeffrey, 57

Kubin, Alfred, 62–63
Kurzweil, Ray, 51, 62

La Loba Loca, 110–111, 138
LaDuke, Winona, 14, 146, 237
Lady Bunny, 36
Lame Deer, 111, 204
land
    colonization of, 89, 160
    Indigenous understandings,
        16, 18, 21, 22, 25, 30, 60, 68,
        70–76, 78, 84, 87, 94, 146, 160,
        194–195
    intelligence of, 91
    language and, 21
    patriarchal understanding
        of, 79, 91
    rape of land, 108, 136
    separation from, 240
"land ethic," 158
"land trauma," 11, 103
language, Indigenous, 18, 114, 151,
    165, 169–170
Lara, Irene, 193
Latour, Bruno, 25
Leach, Edmund, 50
Lenca people, 69–70
Leopold, Aldo, 140, 158,
    164–165, 173
Lerner, Gerda, 77, 79–80
Leslie, Esther, 130, 131
Levy, David, 128
Lickers, Iako'tsi:rareh Amanda, 91
limits, 43, 47, 63, 83, 99, 102,
    125, 129, 132, 136, 139, 140,
    212, 229
Little Bear, Leroy, 18, 68, 174,
    225, 229
Little Mermaid, The (film), 195
Little Shop of Horrors (film), 175,
    179–180
Livingstone, Glenys, 201–202

López-Durán, Fabiola, 118
Lorde, Audre, 32, 184
Love Embrace of the Universe, the
    Earth (Mexico), Me, Diego
    and Señor Xólotol, (Kahlo),
    148–149
Lynas, Mark, 16, 51, 60

maamé, 220–221, 223, 233
Maathai, Wangari, 146
MacKinnon, Catharine A., 89,
    98, 108
Madre Tierra, 39, 66, 193, 195
magic, 152, 161, 175, 182, 202, 235
magic words, 50, 182
Maguire, Gregory, 152
Maleficent, 152
Mamba Mambo (Saar), 156–157
Mambo, 156, 236
"Mammy" stereotype, 122–124
Man, 3, 5, 10, 36, 37, 48, 61, 212,
    214, 230
    desire to copy and replace
        Mother Nature-Earth, 40, 45,
        58, 60, 130–131, 136, 165, 175
    in relation to patriarchal God, 16,
        49, 55, 57
    mass murderous and suicidal, 60
    Sylvia Wynter's conception, 10
Man, The (in African-American
    language), 10, 33, 49, 57, 124,
    183, 217
Man-talk, 19, 21
Mandorla, 225, 234
Mandorla Awakening II: Emerging
    Worlds (Mitchell), 232–234
Maracle, Lee, 87, 174, 190
mass shootings, 15, 88
MasterCard logo, 233
Mawu-Lisa, 30
Mayfield, Curtis, 182
McClintock, Anne, 28–29, 91

McClintock, Barbara, 170
McElya, Micki, 123
McKittrick, Katherine, 89, 128
McLuhan, Marshal, 126, 136
Me Too movement, 32, 103
mechanical bride, 126, 141
mechanical worldview, 47
Medusa, 36, 39, 195, 201, 229
Meikle, Jeffrey, 132, 134–135
menstruation, 17, 128
Merchant, Carolyn, 8, 29,
    46–47, 132
metaphor, 21, 68, 71, 75–76,
    86, 236
Mies, Maria, 10, 30
Miranda, Deborah, 29
misogyny, 4, 28, 29, 31, 50, 88,
    96, 113, 122, 133, 189, 198,
    200, 211
Missy, 10, 20, 65, 76, 184
Mitchell, Nicole, 232–234
Moana (film), 148
Mohawk, John, 24
Momaday, N. Scott, 49, 146, 236
Monáe, Janelle, 35, 182
monsters, 146, 152, 175, 178–180,
    195, 201
    in Navajo understandings, 178–179
Moore, Jason, 102, 146
Mor, Barbara, 26, 30, 94
More, Max, 63
Morrison, Toni, 20, 99, 114, 115,
    116, 159
Morton, Oliver, 145, 148, 156
Moten, Fred, 182
Mother Earth. *See* Mother
    Nature-Earth
Mother Earth Rights, 9, 19, 167
Mother Earth Tree of Life
    (darkdaughta), 196–197
Mother Nature-Earth (Earth
    Mother, Mother Earth, Mother

Nature), 7, 15, 18, 27, 39, 62–63,
    194, 197, 214, 220–222, 225–227,
    233, 234
    in art, 62–63, 148, 229
    defined, 5, 21–22, 213
    colonization of, 66
    gender, 18–19
    in advertising, 12–13, 16–17
    in European thought
    in Indigenous thought, 8
    political art, 66
    sexist stereotypes, 16–17, 44, 46
Mother Who Fucks, 183–184
motherfucker, 3, 27, 78, 82, 109,
    152, 191
    as curse, 40, 203
    as magic, 40, 181–182
    as swear, 142, 183, 203
    culture, 94, 206
    defined, 3, 5, 33–35, 182, 204
    origins, 5, 33, 86
motherfucking, 33, 40, 57, 77, 78,
    89, 108–110, 123, 129, 135, 184,
    199, 229, 236
    as silencing, 198
    defined, 5
    paradigmatic, 108, 166, 197, 229
mothering power, 14, 17, 62, 74, 75,
    119, 126, 144, 191, 193, 197,
    215, 216, 234, 235, 237, 240
mothers and mothering. *See also*
    revolutionary mothering
    African American, 33, 123, 190,
    193, 205–206, 213–214, 217
    as creative power, 62, 185
    dark, 232, 234
    defined, 5, 185, 189, 191, 191–194
    Earth and land as, 60, 68, 94, 185,
    196, 213
    enslaved, 33, 205–206
    essentialist notions, 18
    fear of, 199

mothers and mothering (*cont.*)
  gay, femme and trans
    culture, 193
  Indigenous, 190
  microbial, 171
  Native, 190
  "outraged mother," 221
  phallic separation from, 212
  Pueblo culture, 189, 193–194
  sexist stereotypes, 216, 217
  symbolic, 56, 58, 233
mountains, 13, 48, 118–120, 137–
  138, 164, 172
Mujeres Conciencia (Carrington),
  154–155
"Mutha'"
  as copula, 213
  defined and described, 3, 5
  gender, 5, 183–184, 219, 230,
    234–236
  quotation marks explained, 5–6
Mutu, Wangechi, 229–230
myth, 7, 34, 133, 151, 160, 212, 217
  phallocentric, 57, 61, 99,
    152, 201

Nash, Jennifer, 120
"natural common goods," 70, 99, 19
nature, 88, 111, 112, 158, 227
  Africana conceptions, 25, 159,
    219–223
  agency of, 184, 200
  ancient conceptions, 7, 46–47
  as female, 132
  colonialism and, 66
  defined, 14–15
  distorted conceptions of, 73,
    99, 179
  fear of, 4
  in Onondaga thought, 172
  intelligence and agency of,
    171, 184

Man's quest to dominate, 4, 9–12,
  15, 25, 34, 40, 46–47, 57, 73, 83,
  88, 90–91, 118, 120, 132, 165–
  166, 173, 216
  spirits, 28, 48
  word origin, 6
  worship, 27
"Nature" from *GFP Bunny - Paris
  Intervention* (Kac), 168
Nebula (Prusa), 204, 205, 206
necrotechnology, 61
Nelson, James, 212
neo-necrophilia, 136
Newcomb, Steven, 85
Ngata, Tina, 32
Nixon, Lindsay, 91
#NoDAPL, 68, 69
Nommo, 181, 211
non-binary, 5, 10, 29, 36, 132,
  156, 191
Norbert, Nigit'stil, 111
Nordhaus, Ted, 99
nuke, 92, 94

O'Brien, Ruth, 29
obscenity, 5, 27, 28, 32, 198
omnipotence, 49, 58, 142, 211,
  212, 213
Omolade, Barbara, 86, 104, 106, 108
Opperman, Serpil, 186, 198
organic worldview, 46, 47
Ortiz, Simon J., 21, 146
Oryx and Crake (Atwood), 187
os, 203–204
Osborne, Helen Betty, 97
overwhelm, as synonym for rape, 14
Oya, 37, 228–229

Pachamama, 18, 39, 66
Palin, Sarah, 92
pandemics, 3, 210
panochonas, 20, 203

Patel, Raj, 146
patriarchy, 26, 69, 70, 77–78, 87,
    148, 192, 232
    origins of, 77, 80–81. *See also*
    heteropatriarchy
patterns in nature, 18, 173, 174, 175,
    176, 180, 197, 221, 227, 229
Payne, Katy, 187, 188
penis, 57, 132, 133, 135, 204, 211–212
penis envy, 113
People's Climate March (New York
    City, 2014), 66–68
People of Color's Environmental
    Leadership, 45
People's Climate Defend Our Home
    (Rodriguez), 66, 67
Pérez, Emma, 203
Persephone, 148, 235
Perseus, 195
phallic lust, 28, 29
phallic symbols, 56, 57
phallus, 57, 199, 212
Philip, M. NourbeSe, 82, 105, 106,
    108, 207–208, 211, 228
Phillips, Martin, 78, 88
photosynthesis, 112, 145,
    164–165, 220
"Place-Thought," 94, 95
Plantationocene, 87, 88, 108, 109
plantations, 48, 65, 81, 86, 87, 105–
    109, 123, 207
plastic, 118, 121, 127, 130–141
Plasticocene, 129, 131, 133, 136
plastiglomerate, 129
plastiphobe, 138
plow (plough), 92
Plumwood, Val, 30, 216
popular culture, 4, 5, 15, 33, 78, 83,
    126, 147, 175, 230
porno-tropics, 28–29
pornography, everyday, 29
posthuman, 115

Povinelli, Elizabeth, 11
prayer, 6, 57, 146, 179, 219, 235, 236
Primitif Futur (Duval-Carrié),
    162, 163
principle, the, 206, 208, 210, 222
pronouns, 36
property, 53, 77, 79, 84, 89, 91, 115, 207
Prusa, Carol, 205, 206
Pueblo philosophy, 146
pussy, 5, 32, 206–207, 210–211, 217.
    See also cunt

racism, 4, 31, 45, 69, 70, 72, 76, 79,
    86, 112, 114, 122, 217, 227
rainbow, 30, 36, 112–118, 130–131,
    136–137, 176, 219, 230
rainbow flag, 112
Raine, Nancy Venable, 102
rape, 14, 28, 29, 33, 39, 44, 57, 76–78,
    79, 80–82, 102–103, 123, 129,
    132, 136, 148, 181, 182, 184, 188–
    189, 198, 207, 212, 227, 235, 236
    fracking as, 100–102
    of enslaved Black women, 81, 85–
    87, 106–107, 109, 134, 207–208
    of Native women, 83, 84,
    87, 96–97
    of Nature-Earth, 4, 8, 28, 88, 95,
    100, 110, 136, 179
rape culture, 91, 94, 201
rapism, 4, 140, 198, 212, 236
rationality, 10, 20, 115
reciprocity, 17, 35, 102, 147, 158,
    160, 197
red, 17, 94, 112, 135, 137, 148,
    150–152, 156, 159, 161, 187,
    208, 225
renewal, 3, 8, 14, 17, 120, 126, 141,
    145–147, 173, 216, 219, 230,
    236–237
    in Indigenous philosophy and
    ceremony, 14, 146, 174

reproductive justice, 190, 236
respect, 8, 31, 35, 37, 39, 46, 70,
   95, 99–100, 146, 158, 160,
   171–172, 176, 179, 191–193,
   206, 240
   defined, 208–210
Revkin, Andrew, 24
revolutionary mothering, 20, 85,
   192–193, 196, 197, 237
Richie, Beth, 86
Riley, Shamara Shantu, 10
Rio de Janeiro, 118, 138
ritual. *See* ceremony
rivalry, 8, 13, 14
rivers, 19, 45, 47, 48, 70, 71, 78, 110,
   131, 132, 164
Roach, Catherine, 7
Roberson, Ed, 117
Roberts, Dorothy, 86, 205
Roberts, Jody A., 138, 139
Robinson, Kim Stanley, 137
robots, 61, 115, 121, 127, 128
Rodger, Elliot, 50
Rodriguez, Favianna, 66, 67,
   208, 209
Rodriguez, Juana Maria, 188
Rose, Deborah Bird, 185
Ross, Loretta, 190
Russell, Diana E. H., 14
Russell, Grahame, 70
Rydell, Robert, 122

Saadawi, Nawal el, 150
Saar, Alison, 156, 157
Saar, Betye, 112
Saint Augustine of Hippo, 48, 211
Saint Peter, 51
Salazar, Egla Martínez, 69
Sampson, Kevin, 223, 224, 225, 226,
   227, 233

Sanchez, Carol Lee, 160
Sankofa heart, 197
Santigold, 229, 230
savage, 11, 28, 30, 33, 91, 118, 122, 135
Schellenberger, Michael, 99
Schneemann, Carolee, 203
Schwägerl, Christian, 46, 48
serial killers, 49, 96, 122, 133
Seuss, Dr., 175
Sex Goddess, 125, 134, 148
sex-and-violence, 5, 32, 58–59, 79,
   83, 100, 133, 189
Shange, Ntozake, 128
Sharma, Sarah, 127
Sharpe, Christina, 14, 33
Shaw, Miranda, 150
Shelley, Mary, 175
Shiva, Vandana, 30, 66
Simpson, Leanne Betasamosake, 81,
   84, 147, 160
Singh, Kamaljik, 56
Sjöö, Monica, 26
Sky Woman, 94, 95
skydome, 94
skyscrapers, 51, 55, 56, 117
slavery, 14–15, 30, 79, 104, 107–109,
   115, 122, 127–129, 134, 181,
   205, 207, 220, 227
   in relation to
   Anthropocene, 78, 87
   origins in ancient patriarchy, 77,
   79, 82, 87
Slosson, Edwin, 136, 141
Smith, Andrea, 83
Smith, Linda Tuhiwai, 18, 194, 236
Smitherman, Geneva, 3, 33
Solinger, Ricki, 190
Somé, Malidoma Patrice, 151
Sontag, Susan, 12
Sophia, 151, 156

speciesism, 10, 31
Spencer, Anne, 115
spirit, 23–25, 35–36
"spiritual meanings," 23, 113, 142, 145, 172, 186, 206
Spotted Eagle, Faith, 75, 76
Sprinkle, Annie, 36, 204
stalking, 108
Stallings, L. H., 34, 35, 114
Standing Bear, Luther, 74
Starhawk, 159, 160, 171, 237
Stepford Wives (Levin), 61
Stephens, Elizabeth M., 36
sterility goddess, 141
stories, 185–188
Strayed, Cheryl, 182
Stubbs, Levi, 180
Stuckey, Sterling, 30
Sublette, Constance and Ned, 106
swear, 6, 142, 143, 183, 203
Swentzell, Rina, 98, 193
symbolism, 66, 111
Synthetica, 131, 132, 133, 135, 137
synthetic biology, 60, 61, 109
synthetic rainbow, 113, 130, 136
Szerszynski, Bronislaw, 137

TallBear, Kim, 68, 188
Tampax ads, 17
Tao Teh Ching, 156, 158
Taoism, 229
Tar Baby (Morrison), 116
Tārā, 150
Taylor, Bron, 164
Tea, Michelle, 182
technology, 125, 126, 136, 138, 139, 223, 232
Temple, Christel N., 197
termites, 56

terra nullius, 91, 94, 95, 96, 135
Terra Nullius Is Rape Culture (Lickers and Nelson), 91, 94
territory, 68, 70, 72, 89, 95, 103, 232, 234
Theisen, Olive, 222
Thomas, Dylan, 165
Threadcraft, Shateema, 86
Tiamat, 57
Tinsley, Omise'eke Natasha, 112
Tlaib, Rashida, 82
Todd, Zoe, 20
Torres, Tanya, 161, 162
Trans Day of Resilience Poster (Khadar), 37–38
transgenic art, 166, 167
transhumanism, 167
Transhumanist Manifesto, 64
transmission of affect, 102
transphobia, 31, 82
Tree of Life, 141, 151, 156, 195, 196, 237
trees, 48, 64, 66, 71, 72, 73, 110, 111, 119, 137, 140, 146, 148, 150, 151, 156, 159, 186, 196, 214
Trump, Donald, 82, 120, 207
Trumpocene, 82, 194
Tsing, Anna, 108
Tuama, Pádraig Ó, 21
Tuana, Nancy, 61
Tuck, Eve, 75, 76
Turkle, Sherry, 139
two spirit, 29, 96, 133, 189. See also 2SQ
Tyson, Neil de Grasse, 60

Updike, John, 32
uranium, 178
utopia, 104, 117, 122, 128, 183

Vandermeer, Jeff, 176
Venus, 125–126, 131–136, 141, 150, 153, 165, 184, 228
Violence on the Land, Violence on our Bodies, 97, 103
Virgin Mary, 125, 126, 127
viriditas, 161, 165
Vodou, 112, 156, 162, 236

Walker, Alice, 27, 117, 120, 123, 124, 139, 214–215
Walker, Barbara, 212
Wallace, Michele, 211
Walters, Anna Lee, 46
Wang, K. Wayne, 75
Ward, Jesmyn, 234
Wardlaw, Alvia J., 220
Watts, Vanessa, 30, 47, 71, 72, 73, 94, 95
Web of Life (Biggers), 220, 221, 222
Wells, Ida B., 122
Whanganui iwi Maori, 68
"When Earth Becomes an 'It' " (Awiakta), 2, 37, 144
White Bull, Brenda, 69
"White City," 122, 125, 126
White, Lavinia, 19
White, Lynn, 47
white space, 115–116, 124, 140
White supremacy, 15, 25, 31, 71, 86, 104, 11, 12
whiteness, 84, 114–116, 118, 120–121

'Who's your Daddy," 188–189, 210
"Who's your Mother," 188–189
Williams, Delores, 30, 85, 86, 114
Williams, Mai'a, 192
Wilson, Alex, 18
Winnemucca, Sarah, 84
Wired (magazine), 55, 117, 137, 140
witch, 20, 28, 48, 73, 83, 152–153, 171, 186, 195
Wizard of Oz, The (film), 152
Womack, Ytasha, 223
womb, 33, 47, 98, 105, 106, 107, 124, 125, 129, 132, 141, 195, 213
capitalized, 106, 234
commodified womb, 126
Nature-Earth as, 7, 13, 27, 47, 66, 98, 129, 136, 151, 179, 203, 204, 221, 225, 233–234
womb envy, 13, 31. *See also* cunt envy
wombiverse, 204, 212
Women's March on Washington, 206
Woodman, Marion, 124
Woolf, Virginia, 214
Wynter, Sylvia, 10, 35

Yazzie-Lewis, Esther, 178

Žižek, Slavoj, 46

Printed in the USA
CPSIA information can be obtained
at www.ICGtesting.com
BVHW022150020823
668169BV00001B/1